"We Lost Many Brave Men:"

A Statistical History of the Seventh

Rhode Island Volunteers

By:

Robert Grandchamp

RHODE ISLAND TOWNSHIPS - 1861

"We Lost Many Brave Men"

A Statistical History of the Seventh Rhode Island Volunteers

Robert Grandchamp

HERITAGE BOOKS
2018

HERITAGE BOOKS
AN IMPRINT OF HERITAGE BOOKS, INC.

Books, CDs, and more—Worldwide

For our listing of thousands of titles see our website
at
www.HeritageBooks.com

Published 2018 by
HERITAGE BOOKS, INC.
Publishing Division
5810 Ruatan Street
Berwyn Heights, Md. 20740

Copyright © 2018 Robert Grandchamp

Heritage Books by the author:

From Providence to Fort Hell: Letters from Company K, Seventh Rhode Island Volunteers
"We Lost Many Brave Men": A Statistical History of the Seventh Rhode Island Volunteers
"With their usual ardor", Scituate, Rhode Island and the American Revolution

All rights reserved. No part of this book may be reproduced or transmitted in any form or by any means, electronic or mechanical, including photocopying, recording or by any information storage and retrieval system without written permission from the author, except for the inclusion of brief quotations in a review.

International Standard Book Number
Paperbound: 978-0-7884-5790-6

For Elizabeth

CONTENTS

A brief history of the Seventh Rhode Island Volunteers   1

A new look at Rhode Island Civil War Casualties   7

Introduction   19

Field and Staff   23

Company A   29

Company B   45

Company C   59

Company D   79

Company E   91

Company F   107

Company G   121

Company H   137

Company I   157

Company K   177

Company B (new organization)   195

Company D (new organization)   207

Company G (new organization)   219

Enlistments by town   229

Regimental Statistics	231
Further Reading	247
Acknowledgments	251
About the Author	255

A brief history of the

Seventh Rhode Island Volunteers

On May 22, 1862 Governor William Sprague issued general orders for the raising of the Seventh Rhode Island Volunteers. This regiment was to be the last three-year infantry regiment to be raised in Rhode Island. Camp Bliss was erected in southern Providence and was to be the destination for the recruits for the regiment. Many men came to Camp Bliss in the summer of 1862. A few had seen service in the United States Army and other volunteer regiments. Some were politicians and gentlemen from the hierarchy of the state. The majority were fifteen to thirty-year-old farmers and mill workers from southern and western Rhode Island who enlisted in the regiment under the call of President Abraham Lincoln for 300,000 men to defend the Union following a series of humiliating defeats in Virginia. The largest push for recruits came in August, with some towns offering incentives as high as five hundred dollars for men to enlist. The large bounties encouraged many to come forward to preserve the Union.

The ten companies of the Seventh Rhode Island were recruited primarily from the following Rhode Island communities:

   Company A: Charlestown, Hopkinton, and Richmond
   Company B: Providence
   Company C: Glocester
   Company D: Burrillville and West Greenwich
   Company E: Cumberland, Smithfield, and Woonsocket
   Company F: Exeter and North Kingstown
   Company G: South Kingstown
   Company H: East Greenwich and Warwick
   Company I: Bristol and Newport
   Company K: Coventry, Foster, and Scituate

By the end of August, over 900 Rhode Islanders had gathered at Camp Bliss. To command them, Governor Sprague selected Zenas Randall Bliss of Johnston. Bliss was a graduate of West Point and had attained the rank of captain in the Eighth United States Infantry. In the years ahead he would transform these men from Rhode Island from untrained volunteers into a regiment on par with the United States Regulars.

The Seventh Rhode Island Volunteers was mustered into the service of the United States to serve for three years on September 6, 1862. They also drew Enfield rifle-muskets on this day. In addition, the Seventh was clothed in the full uniform of the United States Army; a feature of their coats being a very high collar. On September 10, the regiment left Rhode Island and proceeded to Camp Casey outside of Washington, D.C. Here they remained for several weeks before joining the First Brigade, Second Division, Ninth Corps on October 6, encamped outside Sharpsburg, Maryland following the victory at Antietam a month earlier. The Seventh remained encamped at Pleasant Valley, Maryland, for three weeks, perfecting its drill, while losing several members of the regiment to a typhoid epidemic that left many ill and unable to perform their duty.

In late October the Army of the Potomac again embarked upon another campaign to capture Richmond, Virginia. In early November, Ninth Corps commander, Major General Ambrose Burnside assumed command of the Army of the Potomac. On November 15, the Seventh fought its first engagement, a brief skirmish in which they held a bridge against Confederate cavalry. Later that month they arrived at Fredericksburg, Virginia. The city lay in their path to attack Richmond. Burnside waited for over two weeks for pontoons to allow his army to cross. The result would be the near destruction of the Seventh Rhode Island.

The Battle of Fredericksburg was one of the worst defeats of the Civil War for the Federal Army. The Army of the Potomac had to attack across a wide open plain to reach a Confederate division entrenched behind a sunken road. In addition, Marye's Heights contained twenty-four pieces of artillery. The Seventh Rhode Island went in at 12:20 on the afternoon of December 13, 1862. Almost immediately, Rhode Islanders were being killed and maimed. Lieutenant Colonel Welcome B. Sayles was hit in the chest by a shell, sprinkling pieces of his body all over members of the Seventh. After halting in the middle of the field to fire their Enfields, the Seventh surged forward in an attempt to capture the wall; they were repulsed by "a perfect volcano of flame," getting to about seventy-five yards from their objective. The Seventh's flag became the farthest advanced banner in the Ninth Corps. After remaining on the field for seven hours, the Seventh was relieved and returned to Fredericksburg. 570 officers and men went into the fight, forty percent became casualties. As the regiment assembled after its charge, all Colonel Bliss could say to his battered regiment was "you have covered yourself with mud and glory." Bliss would be nominated for promotion to brigadier general and receive a Medal of Honor for his actions. Many of his enlisted men would receive promotions for their actions on the field.

Following the Battle of Fredericksburg, the Seventh Rhode Island returned to its winter camp across the Rappahannock River near Falmouth, Virginia. Here many of the men would experience the hardships that their ancestors experienced at Valley Forge some eighty-five years earlier. Food and money was scarce, while typhoid, dysentery, and pneumonia reduced the regiment even further. Even in the worst of weather, one company of the regiment was constantly on picket duty along the river. A respite came in mid-February 1863 when the Ninth Corps was transferred to Suffolk, Virginia. From here they were again transferred to Lexington, Kentucky when Burnside was given command of the Army of the Ohio.

In June the Seventh left Cairo, Illinois as reinforcements for Ulysses S. Grant's army as they besieged Vicksburg, Mississippi. They spent several weeks entrenching around Vicksburg before being sent to Jackson, in order to prevent Confederates from reinforcing the Vicksburg garrison. Here they defeated the Rebels at the Battle of Jackson. Though the Mississippi Campaign only lasted for two months, over fifty Rhode Islanders lost their lives; only three were killed in action. In August, they were recalled to Kentucky. The Seventh entered

Mississippi with slightly over three hundred men; fewer than seventy-five were able to perform duty when the regiment returned to Ohio. Yazoo Fever, dysentery, and typhoid reduced the regiment to mere company strength.

The Seventh spent a miserably cold and wet winter as the provost marshal in Lexington, Kentucky. Here they protected the loyal citizens against John Hunt Morgan's Confederate guerillas. Following this, the regiment was again summoned to Virginia in April 1864 as reinforcements to the Army of the Potomac.

At this time only two hundred and fifty men were on duty. The Seventh returned to Virginia and crossed the Rapidan on the road to Richmond. They were held in reserve at the Battle of the Wilderness on May 5–7, yet were heavily engaged May 12 at Spotsylvania Courthouse, fighting near the Bloody Angle in a driving rain storm. From this day on the Seventh was engaged in combat nearly continuously. The regiment also lost heavily on May 18, 1864 at Spotsylvania, and fought five days later at the North Anna River, where the regimental color bearer was killed.

At Cold Harbor during the first week of June 1864, the Seventh was heavily engaged at Bethesda Church and Mechanicsville, losing over a third of its strength. By this point, Company H had only one man present in line. In mid-June, they arrived at Petersburg with only one hundred and twenty-five men present for duty. As they were constantly under fire, at least one member of the regiment was killed or wounded every day in July and August. On June 20, 1864, only two commissioned officers remained, while most companies mustered ten men, some commanded by corporals. With such a reduced number men, the Seventh Rhode Island was pulled off the line and acted as engineers for the Second Division, Ninth Corps. Colonel Bliss had assumed brigade command after Fredericksburg and never again commanded just the Seventh in battle. After being commanded by several officers during 1864, Captain Percy Daniels was commissioned as lieutenant colonel in June 1864, and commanded the regiment for the rest of the war.

On July 30, at the Battle of the Crater, the Seventh was held in reserve. They remained in their entrenchments throughout August and September, losing even more men. On September 30, 1864, the Battle of Poplar Springs Church was fought and several days later an engagement at Hatcher's Run. In November they were consolidated with the Fourth Rhode Island Volunteers. In addition, men returned to duty and recruits arrived from Rhode Island. By December, over three hundred men were available for duty.

In November the Seventh moved to Fort Sedgwick, also known as Fort Hell as it was the closest fort at Petersburg to the Confederate line. The men lived underground in shelters known as "bombproofs" to escape the murderous fire outside. The Seventh remained here until April 2, 1865, when they helped storm into Petersburg and then pursued Lee to Appomattox Court House. The original regiment was mustered out of the service on June 9, 1864, while the recruits were mustered out on July 13, 1865.

On March 13, 1865, General Ulysses S. Grant formally gave permission for the Seventh Rhode Island Volunteers to paint the following engagements upon their colors where they had fought and died: Fredericksburg, Vicksburg, Jackson, Spotsylvania, North Anna, Cold Harbor, Petersburg, Weldon Railroad, Poplar Spring Church, and Hatcher's Run.

The above was extracted from William P. Hopkins, *The Seventh Rhode Island Volunteers in the Civil War: 1862-1865.* Providence: Snow and Farnum, 1903.

A brief history of the Seventh Rhode Island Volunteers

On May 22, 1862 Governor William Sprague issued general orders for the raising of the Seventh Rhode Island Volunteers. This regiment was to be the last three-year infantry regiment to be raised in Rhode Island. Camp Bliss was erected in southern Providence and was to be the destination for the recruits for the regiment. Many men came to Camp Bliss in the summer of 1862. A few had seen service in the United States Army and other volunteer regiments. Some were politicians and gentlemen from the hierarchy of the state. The majority were fifteen to thirty-year-old farmers and mill workers from southern and western Rhode Island who enlisted in the regiment under the July 1, 1862 call of President Abraham Lincoln for 300,000 men to defend the Union following a series of humiliating defeats in Virginia. The largest push for recruits came in early August, with some towns offering incentives as high as five hundred dollars for men to enlist. The large bounties encouraged many to come forward to preserve the Union.

The ten companies of the Seventh Rhode Island were recruited primarily from the following Rhode Island communities:

Company A: Charlestown, Hopkinton, and Richmond
Company B: Providence
Company C: Glocester
Company D: Burrillville and West Greenwich
Company E: Cumberland, Smithfield, and Woonsocket
Company F: Exeter and North Kingstown
Company G: South Kingstown
Company H: East Greenwich and Warwick
Company I: Bristol and Newport
Company K: Coventry, Foster, and Scituate

By the end of August, over 900 Rhode Islanders had gathered at Camp Bliss. To command them, Governor Sprague selected Zenas Randall Bliss of Johnston. Bliss was a graduate of

West Point and had attained the rank of captain in the Eighth United States Infantry, with active service on the Texas frontier. In the years ahead he would transform these men from Rhode Island from untrained volunteers into a professional fighting force. In time the men of the Seventh would consider themselves on par with the United States Regulars.

The Seventh Rhode Island Volunteers was mustered into the service of the United States to serve for three years on September 6, 1862. They also drew Enfield rifle-muskets on this day. In addition, the Seventh was clothed in the full uniform of the United States Army; a feature of their coats being a very high collar. On September 10, the regiment left Rhode Island and proceeded to Camp Casey outside of Washington, D.C. Here they remained for several weeks before joining the First Brigade, Second Division, Ninth Corps on October 6, encamped outside Sharpsburg, Maryland following the victory at Antietam a month earlier. The Seventh remained encamped at Pleasant Valley, Maryland, for three weeks, perfecting its drill, while losing several members of the regiment to a typhoid epidemic that left many ill and unable to perform their duty.

In late October the Army of the Potomac again embarked upon another campaign to capture Richmond, Virginia. In early November, Ninth Corps commander, Major General Ambrose Burnside assumed command of the Army of the Potomac. On November 15, the Seventh fought its first engagement, a brief skirmish in which they held a bridge against Confederate cavalry. Later that month they arrived at Fredericksburg, Virginia. The city lay in their path to attack Richmond. Burnside waited for over two weeks for pontoons to allow his army to cross. The result would be the near destruction of the Seventh Rhode Island.

The Battle of Fredericksburg was one of the worst defeats of the Civil War for the Federal Army. The Army of the Potomac had to attack across a wide open plain to reach a Confederate division entrenched behind a sunken road. In addition, Marye's Heights contained twenty-four pieces of artillery. The Seventh Rhode Island went in at 12:20 on the afternoon of December 13, 1862. Almost immediately, Rhode Islanders were being killed and maimed. Lieutenant Colonel Welcome B. Sayles was hit in the chest by a shell, sprinkling pieces of his body all over members of the Seventh. After halting in the middle of the field to fire their Enfields, the Seventh surged forward in an attempt to capture the wall; they were repulsed by "a perfect volcano of flame," getting to about seventy-five yards from their objective. The Seventh's flag became the farthest advanced banner in the Ninth Corps. After remaining on the field for seven hours, the Seventh was relieved and returned to Fredericksburg. 570 officers and men went into the fight, forty percent became casualties. It was the greatest loss ever sustained by a Rhode Island regiment in any battle of any war. As the regiment assembled after its charge, all Colonel Bliss could say to his battered regiment was "you have covered yourself with mud and glory." Bliss would be nominated for promotion to brigadier general and receive a Medal of Honor for his actions. Many of his enlisted men would receive promotions for their actions on the field. The regiment received much praise for their actions at Fredericksburg.

Following the Battle of Fredericksburg, the Seventh Rhode Island returned to its winter camp across the Rappahannock River near Falmouth, Virginia. Here many of the men would experience the hardships that their ancestors experienced at Valley Forge some eighty-five years earlier. Food and money was scarce, while typhoid, dysentery, and pneumonia reduced the regiment even further. Even in the worst of weather, one company of the regiment was constantly on picket duty along the river. A respite came in mid-February 1863 when the Ninth Corps was transferred to Suffolk, Virginia. From here they were again transferred to Lexington, Kentucky when Burnside was given command of the Army of the Ohio.

In June the Seventh left Cairo, Illinois as reinforcements for Ulysses S. Grant's army as they besieged Vicksburg, Mississippi. They spent several weeks entrenching around Vicksburg before being sent to Jackson, in order to prevent Confederates from reinforcing the Vicksburg garrison. Here they defeated the Rebels at the Battle of Jackson. Though the Mississippi Campaign only lasted for two months, over fifty Rhode Islanders lost their lives; only three were killed in action. In August, they were recalled to Kentucky. The Seventh entered Mississippi with slightly over three hundred men; fewer than seventy-five were able to perform duty when the regiment returned to Ohio. Yazoo Fever, malaria, dysentery, and typhoid reduced the regiment to mere company strength.

The Seventh spent a miserably cold and wet winter as the provost marshal in Lexington, Kentucky. Here they protected the loyal citizens against John Hunt Morgan's Confederate guerillas. Following this, the regiment was again summoned to Virginia in April 1864 as reinforcements to the Army of the Potomac.

At this time only two hundred and fifty men were on duty. The Seventh returned to Virginia and crossed the Rapidan on the road to Richmond. They were held in reserve, but came under fire at the Battle of the Wilderness on May 5–7. The regiment was heavily engaged May 12 at Spotsylvania Court House, fighting near the Bloody Angle in a driving rain storm. From this day on the Seventh was engaged in combat nearly continuously. The regiment also lost heavily on May 18, 1864 at Spotsylvania, and fought five days later at the North Anna River, where the regimental color bearer was killed.

At Cold Harbor during the first week of June 1864, the Seventh was heavily engaged at Bethesda Church and Mechanicsville. Over a third of the regimental strength was lost in fighting on June 3. By this point, Company H had only one man present in line. In mid-June, they arrived at Petersburg with only one hundred and twenty-five men present for duty. As they were constantly under fire, at least one member of the regiment was killed or wounded every day in July and August. On June 20, 1864, only two commissioned officers remained, while most companies mustered ten men, some commanded by corporals. With such a reduced number of men, the Seventh Rhode Island was pulled off the line and acted as engineers for the Second Division, Ninth Corps. Colonel Bliss had assumed brigade command after Fredericksburg and never again commanded the Seventh in battle. After being commanded by several officers during 1864, Captain Percy Daniels was commissioned as lieutenant colonel in June 1864, and commanded the regiment for the rest of the war.

On July 30, at the Battle of the Crater, the Seventh was held in reserve. They remained in their entrenchments throughout August and September, losing even more men. On September 30, 1864, the Battle of Poplar Springs Church was fought and several days later an engagement at Hatcher's Run. In November they were consolidated with the Fourth Rhode Island Volunteers. In addition, men returned to duty and recruits arrived from Rhode Island. By December, over three hundred men were available for duty.

In November the Seventh moved to Fort Sedgwick, also known as Fort Hell as it was the closest fort at Petersburg to the Confederate line. The men lived underground in shelters known as "bombproofs" to escape the murderous fire outside. The Seventh remained here until April 2, 1865, when they assisted in the storming of Petersburg, and then pursued Lee to Appomattox Court House. The original regiment was mustered out of the service on June 9, 1864, while the recruits were mustered out on July 13, 1865.

On March 13, 1865, General Ulysses S. Grant formally gave permission for the Seventh Rhode Island Volunteers to paint the following engagements upon their colors where they had fought and died: Fredericksburg, Vicksburg, Jackson, Spotsylvania, North Anna, Cold Harbor, Petersburg, Weldon Railroad, Poplar Spring Church, and Hatcher's Run.

The above was extracted from William P. Hopkins, *The Seventh Rhode Island Volunteers in the Civil War: 1862-1865*. Providence: Snow and Farnum, 1903.

A new look at Rhode Island Civil War Casualties

Civil War historians have long quoted that 620,000 American soldiers, North and South died in the Civil War or as a result of their service. In 2012, Dr. David Hacker of Binghamton University, using the latest available data, stunned the Civil War community by announcing that the casualty figure is actually much higher, nearly 750,000 Union and Confederate military dead. Based on census data, a careful look at the casualty rates among black and immigrant soldiers, and a review of filed pension applications; Dr. Hacker's figure is widely gaining ground in the field as the true number of men who died as a result of their service. For the record, this historian agrees with Dr. Hacker's figure, however, the true number will never be known.[1]

In my research on Rhode Island's role in the Civil War, I have long had a nagging suspicion that the state lost far more men than originally claimed. According to Lieutenant Colonel William F. Fox in his massive *Regimental Losses in the American Civil War,* Rhode Island furnished 23,236 men to serve in the war. It is important to note that this figure includes all Rhode Island enlistments and not just those native-born Rhode Islanders who signed up; many men, most notably those who served in the Fourteenth Rhode Island as well as the Second Rhode Island Cavalry came from other states. Of these, according to Fox, 460 men were killed in action or mortally wounded and 861 died "from all other causes," including of disease, as prisoners of war, and in accidents for a total of 1,321 military deaths during the war. In his book, Fox offered a stern warning to future scholars regarding Civil War casualty figures. "Days, and often weeks, have been spent on the figures. It is hoped that before disputing any essential fact, a like careful examination of the records will be made." Despite Colonel Fox's statements, it is worth revisiting Rhode Island's Civil War casualty figures.[2]

---

[1] *New York Times,* April 2, 2012.
[2] William F. Fox, *Regimental Losses in the American Civil War: 1861-1865.* (Albany: Albany Publishing Company, 1889), preface, 526, 533.

In his 1964 book, *History of the Rhode Island Combat Units in the Civil War,* General Harold Barker, a veteran of the First and Second World Wars, whose grandfather had served in the Civil War recorded a total of 1,685 men from Rhode Island units who died as a result of their Civil War service. Because he did not footnote his book, it is unclear how General Barker reached this conclusion.[3]

Immediately after the Civil War, the Rhode Island General Assembly appointed a committee of prominent Rhode Islanders, including Ambrose Burnside and John Russell Bartlett to find and accept a proposal for a statewide monument that would list the names of every Rhode Islander who died in the "wicked Rebellion." The monument, officially the Rhode Island Soldiers' and Sailors' Monument, would be inscribed, "Rhode Island pays tribute to the memory of the brave men who died that their country might live." After a year-long search, the committee settled on a design from Randolph Rogers, consisting of a statue of "America Militant," four bronze panels representing War, Victory, Peace, and History, as well as four additional figures representing the infantry, cavalry, artillery, and navy. Most importantly were the twelve panels that would contain the names of every Rhode Islander who died in the war. The entire monument cost 50,000 dollars and was dedicated in Exchange Place (Kennedy Plaza) on September 16, 1871 to much fanfare.[4]

The first step in the monument process was carefully going through the records held by the adjutant general of Rhode Island and compiling a listing of the names to be inscribed on the bronze panels. This in and of itself was an arduous task. When he inherited the records in the early 1890's Elisha Dyer Jr., himself a Civil War veteran and the adjutant general complained about the terrible condition of Rhode Island's Civil War records, "The old

---

[3] Harold R. Barker, *History of the Rhode Island Combat Units in the Civil War: 1861-1865.* (Providence: NP, 1964), 307-308.
[4] For more on the Rhode Island Soldiers' and Sailors' Monument, refer to *Report on the Committee on a Monument to the Rhode Island Soldiers' and Sailors' who perished in suppressing the Rebellion Made to the General Assembly, January Session, 1867.* (Providence: Providence Press, 1867) and *Proceedings at the Dedication of the Soldiers' and Sailors' Monument in Providence.* (Providence: A. Crawford Greene, 1871)

and valuable records of the Rhode Island Regiments were being irreparably injured by the constant handling of those who were obliged to refer to them for information. From the close of the war until June, 1883, the records were kept in paste-board boxes in an open bookcase in the Adjutant-General's office, where they were easily accessible to the public, and, consequently, also in danger of being carried off and lost, as well as being spoiled or destroyed by careless handling."[5]

The committee only had a year to go through the thousands of muster rolls, as well as compiling a listing of men from Rhode Island who had served in the Regular service or in units from other states. Despite this momentous task, the committee completed its task and recorded the names of 1,771 Rhode Islanders who died of wounds, disease, in prisons, of accidents in the service, or of disease immediately upon returning home from the army. These 1,771 names were inscribed upon the monument in Providence. Many names were missed however.[6]

So, who is to be believed? Fox widely regarded as the leading authority on Civil War statistics, or the adjutant general's office in Rhode Island, who was responsible for recording the deaths of Rhode Island's soldiers and sailors. The discrepancy between Fox's figures and those of the state are 450. When one deducts Rhode Islanders who served in the units of other states, as well as in the U.S. Army, U.S. Navy, and Marine Corps, Rhode Island still records a figure of 358 military deaths in Rhode Island

---

[5] Elisha Dyer, *Annual Report of the Adjutant General of the State of Rhode Island and Providence Plantations for the Year 1865. Corrected, Revised, and Republished in Accordance with the Provisions of Chapters 705 and 767 of the Public Laws. Volume I* (Providence: E.L. Freeman, 1893), i. This book, the official listing of Rhode Island's Civil War soldiers is more commonly referred to as *The Revised Register of Rhode Island Volunteers.* Volume One covers those who served in the infantry, while Volume Two covers the cavalry, artillery, Regulars, and U.S. Navy. In the mid-1990s Kenneth Carlson of the Rhode Island State Archives began a meticulous project to finally catalog all of the Civil War papers from Rhode Island. Today, the papers, at the Rhode Island State Archives contain the best records of Rhode Island's Civil War soldiers.

[6] *Names of the Officers, Soldiers, and Seamen in Rhode Island Regiment, or Belonging to the State of Rhode Island, and Serving in the Regiments of other States and in the Regular Army and Navy of the United States, who lost their live in the Defence of their Country in the Suppression of the Late Rebellion.* (Providence: Providence Press, 1869)

regiments over the usually quoted number given by Fox. The question is who is to be believed. In the opinion of this historian, the higher number, quoted by the State of Rhode Island in the 1869, the very names that were inscribed on the Soldiers' and Sailors' Monument are the correct figure. But a deeper question remains, is the figure of 1,771 Rhode Islanders dying as a result of their Civil War service an accurate figure. Could the number be higher?

I have long held suspicion that the number of Rhode Islanders who died as a result of their Civil War service was over 2,000 and I regularly quoted the number in my books and lectures on the Civil War. In 2014, I set out to test my hypothesis, to finally determine the number of Rhode Islanders who died in the Civil War.

I began my study by focusing on the Seventh Rhode Island Volunteers, a regiment that based on my great-great-great uncle Alfred Sheldon Knight's service, I was infinitely familiar with. According to Fox, the Seventh sustained a loss of ninety officers and men killed in action and died of wounds, as well as 109 who died "died of disease, accidents, in prison &c." In his only other reference to the regiment, Fox stated that at the Battle of Fredericksburg the Seventh sustained a loss of eleven dead, 132 wounded and fifteen missing for a total of 158. While acknowledging the limited resources of muster rolls and after action reports that Fox worked with, I knew these figures were woefully low.[7]

Based on a survey of records which have included entries in soldier's letters and diaries, cemetery records, pension files, town clerk death listings; pretty much every scrap of paper that exists regarding the Seventh Rhode Island Volunteers, I determined that the Seventh, which carried a total of 1,179 men on its roles during the war sustained a loss of 104 officers and men who died in combat or of wounds sustained in battle, as well as 109 who died of other causes such as disease and accidents.

---

[7] Fox, *Regimental Losses,* 434:473.

Not included in my figures are the twenty-seven men from the Seventh who died after being mustered out of the army, but whose deaths are directly attributable to their Civil War service. Among them is Lieutenant Colonel Job Arnold of Providence. He was discharged for disability in May 1864 after contracting malaria in Mississippi. Arnold died in Providence in December of 1869, and as reported in local papers, he died as a direct result of his service in the Seventh Rhode Island in the Deep South.[8]

Countering Fox's claim regarding the Seventh's losses at Fredericksburg, my determination is the regiment lost three officers and forty-six men killed in action or mortally wounded, as well as 145 officers and men wounded; in addition, three men were captured. This is a total of 197 officers and men of the roughly 570 who went into the fight.[9]

Because of the gruesome nature of Civil War combat, where minie balls and exploding artillery rounds left mangled corpses scattered around the ground, many men were simply listed as missing in action or deserted, leaving families searching, in some cases for years, as to what happened to their family member. On September 17, 1862, the Fourth Rhode Island advanced into Otto's Cornfield at the Battle of Antietam. Forming the extreme left flank of the Army of the Potomac, the Rhode Islanders were flanked in the dense corn and disintegrated under fire. According to one veteran from the regiment, "our men fell like sheep at the slaughter." After a few disjointed volleys, the men of the Fourth

---

[8] *Manufacturers and Farmers' Journal,* January 3, 1870 and *Newport Mercury,* January 1, 1870. William P. Hopkins, *The Seventh Regiment Rhode Island Volunteers in the Civil War: 1862-1865.* (Providence: Snow & Farnum, 1903), 322. As stated above, my recording of casualty figures has been exhaustively researched from all available soldiers' letters and journal entries, pension and service files at the National Archives, death listings in town hall records, as well as cemetery visits.
[9] Zenas Randall Bliss, *The Reminiscences of Major General Zenas R. Bliss: 1854-1876.* Edited by Thomas T. Smith, Jerry D. Thompson, Robert Wooster, and Ben E. Pingenot. (Austin: Texas State Historical Association, 2007), 324-330. Hopkins, *Seventh Rhode Island,* 47-59. Company A, Seventh Rhode Island Monthly Returns, December 1862, Author's Collection. William P. Hopkins gave the Seventh's casualty figures as thirty-nine dead and 120 wounded for a total loss of 159. My figure includes men who later died of wounds and those whose injuries were recorded in myriad of sources including newspapers, letters, journals, and pension records.

fled for their lives. According to the official regimental report field five days after the battle by Lieutenant Colonel Joseph B. Curtis, the regiment lost twenty-one killed, seventy-seven wounded, and two men missing in action. These figures have long been substantiated as the toll of battle for the Fourth on that terrible day.[10]

One of the Fourth's veterans who fell that day was Corporal Austin A. Perkins of Richmond, Rhode Island. According to veteran George H. Allen in his 1887 book *Forty-Six Months in the Fourth Rhode Island Volunteers,* Perkins "deserted at Antietam." In a copy of Allen's book held in the archives of Providence College are numerous margin notes obviously written by a veteran of the Fourth who double checked all of Allen's statistics. In the entry for Corporal Perkins he wrote, "Later believed to have been killed at Antietam Sept. 17, 1862." Fortunately, in the early 1890s, as Elisha Dyer prepared to publish his massive *Revised Register,* it was noted under the entry for Austin "Believed to have been killed in the battle of Antietam." Despite the official nod from the state that this soldier lost his life in combat, as in many cases, no additional names were ever added to the Soldiers and Sailors Monument, as such the name of Corporal Austin A. Perkins is not listed on the monument in Providence. With the hindsight of a century and a half to carefully check all the records, it is now believed that the true casualty count of the Fourth in Otto's Cornfield, including those who later died of wounds was thirty dead seventy-one wounded, and four captured. Furthermore, nineteen men are listed as having "deserted in the face of the enemy." Going into the battle with 221 men, the regiment lost well over half of its strength, percentage wise the most ever lost by a Rhode Island unit in any battle of any war.[11]

If the casualty figures can be so different for just the Seventh Rhode Island, I knew that they would increase as well for the other regiments Rhode Island sent to the war. To begin my

---

[10] Joseph B. Curtis to William Sprague, September 22, 1862, Rhode Island State Archives. *Providence Journal,* September 25, 1862.
[11] George H. Allen, *Forty-Six Months in the Fourth Rhode Island Volunteers.* (Providence: J.A. & R.A. Reid, 1887), 371-389. *Revised Register: Volume One,* 301. *Proceedings at the Dedication,* 9-10. John Michael Priest, *Antietam: The Soldiers Battle.* (Oxford: Oxford University Press, 1994), 277-278: 351.

research, I drew up a plan to visit every town hall in Rhode Island, in addition to combining with a search through historic cemeteries. I gave myself very limited parameters. The notation of death entered into the ledger by the clerk had to clearly indicate that the man died as a direct result of his Civil War service. In the occupation field, the man would be listed as a "soldier," "volunteer" or "in U.S. Service." Under the heading of death, the notation had to clearly indicate that the man died of wounds or disease he encountered in the army. For example, I encountered in the Cranston records a recently discharged soldier who was run over by a railroad car shortly after returning home; this would not qualify as a Civil War death. In the graveyard search, the inscription on the headstone had to clearly indicate a Civil War death, such as the often encountered "Died of disease contracted in the service of his country" or "Died of wounds received in the Battle of..."

Although the Rhode Island General Assembly had required city and town clerks to record births, marriages, and deaths at the local level beginning in 1853, and had even sent books to the clerk offices for this purpose, by the 1860s, the system was still woefully inaccurate. Many such vital records continued to only be recorded in family Bibles. Some clerks such as those in Providence, Scituate, Coventry, and Warwick maintained meticulous records from the time, recording the deaths of men who were residents of the town, but died out of state while on military service, as well as those who died of wounds or disease at home. Indeed, the city clerk in Providence even took the time to record the street and address of where the deceased died. In addition, when a soldier died, he listed their unit. Surprisingly towns such as Burrillville, Glocester, Little Compton, and Westerly, who lost soldiers in the Civil War, and whose death records can be found elsewhere recorded few soldiers between 1861-1865 in their town vital records.[12]

The quest to determine the number of Rhode Islanders who died in the Civil War, specifically those who came home and died of wounds or illness took me to every clerk's office in Rhode

---

[12] Death Records for Burrillville, Coventry, Glocester, Little Compton, Providence, Scituate, Warwick, Westerly, contained in the clerk's offices in those communities.

Island. During the course of my research into these records, to date, I recorded that the clerks had recorded over 400 soldiers in their records who died in the war; the vast majority of whom died at home of wounds or illness contracted in the service, not in camp or on the battlefield. A good example is Samuel Towne of Battery C, First Rhode Island Light Artillery. He died in North Providence on February 13, 1863 of dysentery "contracted in Chickahominy."[13]

It will take years to wade through the records recorded in the search, to cross check names against pension files, as well as those recorded in the *Revised Register of Rhode Island Volunteers*. While the research continues, here are several examples of what this research has uncovered thus far.

Alpheus Salisbury was a married, thirty-year-old weaver from Scituate who served in Company K, Seventh Rhode Island Volunteers. Salisbury was shot in the neck in the Seventh's assault up Marye's Heights at the Battle of Fredericksburg. He was discharged from the service for disability on February 2, 1863 and sent home to Scituate. According to a published medical report filed by local doctor William H. Bowen who treated Salisbury:

> The most prominent symptoms were great pain in the head, frequent vomitings, constipation, and a kind of stupor. The wound in the head had not healed, and on probing it pus and blood were discharged. He learned that several pieces of bone had been taken away since the injury was inflicted. On July 1st, he saw the patient, in consultation with another physician. Pain in the head and vomiting still continued, and there was more perfect unconsciousness. The next morning there was paralysis of the side opposite the wound in the head, with one pupil contracted while the other was dilated, and he was perfectly comatose. He thinks that the wound was the primary and the original cause of death.[14]

---

[13] Register of Death entry for Samuel Towne, 1863 North Providence Death Returns, Pawtucket City Hall, Pawtucket, RI.
[14] Joseph K. Barnes, *The Medical and Surgical History of the War of the Rebellion*. (Washington, DC: Government Printing Office, 1875), 201

Private Salisbury died on July 2, 1863 as a direct result of his injuries sustained at Fredericksburg some seven months earlier; he was buried in the Clayville Cemetery. Federal pension clerks agreed with Dr. Bowen's findings and granted his wife a pension based on the fact that he died of his injuries sustained in government service. Despite the government's findings, the name of Alpheus Salisbury is not recorded on the Soldiers' and Sailors' Monument in Providence.[15]

Some Rhode Island families lost two of their sons in the service of the Union; one Foster family lost three. Among those who lost two was the Pearce family of Richmond. William and Harvey Pearce enlisted in Battery B, First Rhode Island Light Artillery in March 1862. William quickly fell ill on the Virginia Peninsula and was discharged for disability and sent home on June 30, 1862; Harvey struggled on until March 20, 1863, when he too was discharged for disability. Both men returned to Hopkinton where, according to inscriptions on both headstones, they "died of disease contracted in the U.S. Service during the Great Rebellion." The two Pearce brothers are buried side by side at Wood River Cemetery in Hope Valley, the only indication they died in the service is the inscription upon their now fading headstones. Neither name is inscribed on the monument in Providence, or in the local town records.[16]

The Seventh Squadron of Rhode Island Cavalry is one of the state's more interesting Civil War units. Composed of one company raised from students from Dartmouth and Norwich, and one from men from northern Rhode Island, the squadron spent an uneventful three months of service in the Shenandoah Valley in the summer of 1862. Indeed, the regiment's only glory came in the last days of their enlistment when they participated in a wild breakout from the Harpers Ferry garrison. According to the army records, only one squadron member, Arthur Coombs of Thetford, VT, a student from Norwich died of typhoid when the squadron was stationed near Winchester, Virginia. Another Seventh Squadron casualty was Henry C. Colwell of Glocester. He died of

---

[15] Alpheus Salisbury, Pension File, National Archives. *Proceedings at the Dedication*, 55.
[16] Harvey and William Pearce Headstones, Wood River Cemetery, Richmond, RI. *Revised Register: Volume II*, 784. *Proceedings at the Dedication*, 61-62.

typhoid on November 3, 1862 in Chepachet, a month after returning home from the front. While recorded in the town clerk's records, Colwell's name is not on the Soldiers' and Sailors' Monument, nor is that of Private Coombs.[17]

Ira E. Cole was seventeen and a farmer from Foster when he enlisted in Company E of the Third Rhode Island Heavy Artillery in the summer of 1861. Assigned to artillery duty in Georgia, Florida, and South Carolina, the Third, like the vast majority of Civil War regiments lost far more men to illness than to the enemy's guns. Private Cole survived his three-year enlistment unscathed. He returned to Foster in the summer of 1864 a sick man however. On August 31, 1865, according to the town clerk's notations, Cole died of "chronic dysentery contracted in camp" at his home in Foster. Like the majority of the men chronicled in this study, his name was not listed as a Civil War death of state authorities.[18]

Perhaps the most interesting find so far in the search has been a remarkable discovery regarding Private Ira Cornell of Coventry, who served in Company K of the Seventh Rhode Island Volunteers. Cornell was a farmer from Coventry who enlisted on August 14, 1862. A day later, his son Ira Cornell Jr. also enlisted. The senior Cornell was wounded at Fredericksburg. According to the official army records, he "deserted at Cincinnati, O. April 1, 1863." His son, Ira Cornell junior was discharged for disability on October 14, 1864 and died of tuberculosis contracted in the army on April 29, 1867 in Coventry.[19]

When I conducted my initial survey of compiling a roster of the Seventh Rhode Island, I listed the senior Cornell as did the official records, as having deserted. In the back of my mind however, was why a father would desert the army, leaving his

---

[17] Register of Death entry for Henry C. Colwell, 1862 Death Returns, Glocester Town Hall, Chepachet, RI. *Revised Register: Volume II,* 305-310. *Proceedings at the Dedication,* 63-65.
[18] Frederic Denison, *Shot and Shell: The Third Rhode Island Heavy Artillery Regiment in the Rebellion, 1861-1865.* (Providence: J.A. & R.A. Reid, 1879) Register of Death entry for Ira E. Cole, 1865 Death Returns, Foster Town Hall, Foster, RI. *Proceedings at the Dedication,* 50.
[19] Hopkins, *Seventh Rhode Island,* 522. Register of Death entry for Ira Cornell Jr., 1867 Death Returns, Coventry Town Hall, Coventry, RI.

teenage son alone in the service. One day this past fall while researching in the Coventry Town Hall, I discovered a notation that literally blew me away. The town clerk in Coventry kept meticulous records of all soldiers who died in the army from the town. As I was busy looking through the register of deaths in Coventry, I saw the name of Ira Cornell. In the margin, the clerk had annotated, "Drowned in the Ohio River in the attempt of crossing it in the line of his duty." This record makes sense, for on that date the steamer *Kentucky,* transferred the Seventh Rhode Island across the Ohio River from Cincinnati to Covington, Kentucky. I have found no reference of the death in any letters from Seventh Rhode Island soldiers recording this, nor is it recorded in the official history of the regiment. Despite this, the Coventry records are highly accurate, and the clerk would have received first-hand information from a fellow soldier or a relative about the death. In addition, there is a grave marker in Pine Grove Cemetery in Coventry for Cornell that gives a date of death of April 1, 1863. In my records of the Seventh Rhode Island I have changed my entry to reflect that Ira Cornell died in the service of his country and did not desert the flag.[20]

Will the number of Rhode Island soldiers who died as a result of battle injuries or of disease contracted in the service ever be known? The answer is probably not. Men immediately left the state after mustering out, some never returned, many never had their deaths recorded in the vital records, and today lie buried in an unmarked grave. The best available data supports that approximately 2,000 Rhode Island soldiers, sailors, and Marines died as a result of their Civil War service, which severely contradicts the previous numbers as being too low.

Perhaps Thomas Williams Bicknell summed it up best in his massive *The History of the State of Rhode Island and Providence Plantations* when describing the Civil War registers from the smallest state: "This report gives the name, date of enrollment, dates of mustering in and mustering out, promotions, transfers, or all soldiers and sailors from Rhode Island in the Civil War. Totals are not given and no record as to the nationality, birth,

---

[20] Register of Death entry for Ira Cornell, April 1, 1863, Coventry Town Hall. Hopkins, *Seventh Rhode Island,* 69-70.

or birth-place of any of the whole number. In most cases the Rhode Island residence is noted. From these data, it is almost impossible to determine how many men Rhode Island contributed to the War. It can safely be stated that the State furnished the full quota of men and supplies that she was called to render."[21]

---

[21] Thomas Williams Bicknell, *The History of the State of Rhode Island and Providence Plantations: Volume II.* (New York: American Historical Society), 820-821.

INTRODUCTION

The following roster of the Seventh Rhode Island Volunteers was carefully transcribed from the original muster rolls and descriptive books held at the Rhode Island State Archives. The roster lists all the men known to have actually served in the regiment from 1862-1865, and does not include those who deserted from Camp Bliss before the regiment was mustered in on September 6, 1862, or those men who never actually appeared in camp after they were transferred from the Fourth Rhode Island Volunteers in the fall of 1864. In certain cases additional information regarding casualties, deserters, and those who died at home has been added to the register from sources including Rhode Island newspapers, town hall records, the letters and journals of members of the Seventh Rhode Island, pension and service records, as well as the personal observations by this writer in the cemeteries of Rhode Island and elsewhere.

It is important to note the men who died of disease or wounds after they were mustered out or discharged from the service due to disability and are not listed elsewhere as officially dying in the service; these men died as a direct result of their military service and are recorded here as such.

Although some information might not be supported by what is listed in the "official records," this roster represents the most complete and accurate set of data of the officers and men who served in the Seventh Rhode Island Volunteers. Furthermore, when known, the date of death and burial location of the veteran has been identified. To identify burial locations, each soldier was run through three databases, namely www.findagrave.com, the Rhode Island Historical Cemetery Database, as well as Civil War veteran burial locations of the Sons of Union Veterans of the Civil War. Furthermore, the Veterans Affairs National Cemetery database was also utilized. No doubt, many of these men who were killed in action or who died of illness in the South will forever rest in a National Cemetery under a stone marked "Unknown." A cenotaph is a memorial marker in a cemetery in Rhode Island, the soldier's body remains buried in the South.

The result of this intense research was the location and identification of the burial locations of nearly two thirds of the men who served in the regiment. In addition, historic cemetery records located in each town hall in Rhode Island were also utilized. Furthermore, this author has personally visited nearly every cemetery in Rhode Island and has visited many of the graves recorded in these pages, confirming the identity of the man buried there as a member of the Seventh Rhode Island Volunteers.

Each man is identified, followed by the residence he claimed to hail from on the enlistment form. His age upon enlistment is also given, as is his marital status upon enlistment, designated by an "S" for single or "M" for married. Furthermore, his occupation is noted. Recruited from all over Rhode Island, the Seventh represented every walk of life in Civil War era New England. In addition, significant milestones in the soldier's life such as when and how they died or were injured in action, died of disease, discharged for disability, mustered out, or transferred to other organizations is noted.

In regards to the residence of the soldier, the residence is that claimed upon the enlistment form, or when not listed, where the soldier last knowingly resided. For example, Colonel Zenas Randall Bliss was a native of Johnston, but had not resided in the town since 1850 when he went to West Point, graduating in 1854 and then spending the next seven years on the Texas frontier. He is listed in this roster as a resident of Johnston. In the case of Woonsocket, it should be noted for the record that Woonsocket did not officially exist as a separate town until 1867. In 1862, the village of Woonsocket consisted of a large industrial area comprising both the towns of Smithfield on the west bank of the Blackstone River, and Cumberland on the east side. It is nearly impossible to determine which side of the river a soldier may have resided on. As many soldiers claimed the then unincorporated village of Woonsocket as their home, it is listed in this roster as a separate place of residence then Cumberland and Smithfield. Other Rhode Island boundaries have changed over time as well; readers are advised to consult the wonderful book *Rhode Island Boundaries: 1636-1936* by John Hutchins Cady.

In the fall of 1864, the severely depleted ranks of the Seventh Rhode Island were consolidated from ten companies into seven. Members of Companies B, D, and G of the regiment were reassigned to the Seventh's other companies. With the mustering out of the Fourth Rhode Island Volunteers in October of 1864, the veterans and recruits of the Fourth were consolidated into three companies and reassigned to the Seventh Rhode Island as Companies B, D, and G; these are sometimes referred to as the New Organization.

This register is the result of over fifteen years and *countless* hours of study and revision. It will stand the test of time as the most complete and accurate record of a Rhode Island regiment during the Civil War. The men who served in the Seventh Rhode Island Volunteers did their duty to free the slave and preserve the Republic. I hope I have done their memory justice in preserving an accurate record of their name and deeds for posterity.

ROBERT GRANDCHAMP
JERICHO CENTER, VERMONT

FIELD AND STAFF

Colonel

Bliss, Zenas Randall. Residence, Johnston. 29. S. Soldier. Commissioned Sept. 4, 1862. Mustered in September 6, 1862. Slightly wounded in arm at Fredericksburg, VA, Dec. 13, 1862. Brevet major U.S. Army for gallantry at Fredericksburg, VA, Dec. 13, 1862. Awarded the Medal of Honor for heroism at Fredericksburg, VA. Brevet lieutenant colonel U.S. Army for gallantry at the Battle of the Wilderness, May 5-7, 1864. Ordered on detached service at Wheeling, WV, Nov. 9, 1864, and so borne until June 1865. Mustered out of volunteer service June 28, 1865 and returned to the rank of captain and brevet lieutenant colonel, United States Army, and sent to next duty station at Philadelphia, PA. Died Jan. 1, 1900. Interred at Arlington National Cemetery. Arlington, VA.

Lieutenant Colonels

Arnold, Job. Residence, Providence. 35. S. Engraver. Transferred from Fifth RI Heavy Artillery, Mar. 2, 1863. Discharged for disability May 28, 1864. Died of malaria contracted in the service at Providence, RI, Dec. 28, 1869. Interred at North Burial Ground, Providence, RI.

Church, George E. Promoted from Capt. Co. C, Jan. 7, 1863. Promoted to colonel Eleventh RI Vols. Feb. 11, 1863. Died Jan. 4, 1910. Interred at Swan Point Cemetery, Providence, RI.

Daniels, Percy. Promoted from Capt. Co. E, June 29, 1864. In command of regiment from May 18, 1864, until its muster out. Brevet colonel July 30, 1864 for actions at the Crater, Petersburg, VA. Mustered out June 9, 1865. Died Feb. 14, 1916. Interred at Girard Cemetery, Girard, KS.

Sayles, Welcome Ballou. Residence, Providence. 50. M. Printer. Commissioned May 22, 1862. Mustered in September 6, 1862.

Killed in action at Fredericksburg, VA, Dec. 13, 1862. Interred at Swan Point Cemetery, Providence, RI.

Majors

Babbitt, Jacob. Residence, Bristol. 53. M. Banker. Commissioned Sept. 4, 1862. Mustered in September 6, 1862. Mortally wounded in action, shot in chest, at Fredericksburg, VA, Dec. 13, 1862. Died of wounds at Alexandria, VA, Dec. 23, 1862. Interred at Juniper Hill Cemetery, Bristol, RI.

Jenks, Ethan Amos. Promoted from captain Co. I, June 29, 1864. Never mustered in as such, served as major for duration of service. Wounded in action, in shoulder, at Petersburg, VA, Mar. 20, 1865. Mustered out June 9, 1865. Died May 13, 1901. Interred at Tourtellot Family Lot, Johnston Cemetery 22, Johnston, RI.

Tobey, Thomas F. Promoted from captain Co. E, Jan. 7, 1863. Discharged for disability Feb. 9, 1864. Died June 7, 1920. Interred at North Burial Ground, Providence, RI.

Adjutants

Page, Charles F. Residence, Bristol. 23. S. Book keeper. Commissioned Sept. 4, 1862. Mustered in September 6, 1862. Wounded in action, shot in head and lost left eye, at Fredericksburg, VA, Dec. 13, 1862. Discharged for disability Feb. 23, 1863. Died Oct. 6, 1891. Interred at Swan Point Cemetery, Providence, RI.

Sullivan, John. Promoted from second lieutenant Co. K, Mar. 1, 1863. Captured at Jackson, MS, July 13, 1863. Paroled at James River, VA, Feb. 22, 1865. Mustered out June 9, 1865. Died July 3, 1872. Interred at Loveland, CO.

Spooner, Henry Joshua. Residence, Providence. 23. S. Lawyer. Commissioned Oct. 15, 1862. Transferred from 4th RI Vols. Oct. 15, 1864. Mustered out April 15, 1865. Died Feb. 9, 1918. Interred at Swan Point Cemetery, Providence, RI.

Regimental Quartermasters

Fessenden, Samuel. Promoted from sergeant major to second lieutenant and regimental quartermaster Oct. 20, 1863. Promoted to first lieutenant and quartermaster Nov. 13, 1863. Dismissed from the service for embezzlement Dec. 13, 1864. Died Feb. 11, 1894. Interred at Moshassuck Cemetery, Central Falls, RI.

Linnell, Dean S. Residence, Pawtucket. 42. S. Machinist. Transferred from 10th R.I. Vols. Aug. 1862. Resigned Nov. 3, 1862. Died Oct. 17, 1867. Interred at Oak Grove Cemetery, Pawtucket, RI.

Morse, Ephraim C. Transferred from first lieutenant Co. G, Jan. 11, 1865. Mustered out at Providence July 25, 1865. Died July 30, 1885. Interred at Oak Hill Cemetery, Auburn, ME.

Stanhope, John R. Jr. Promoted from quartermaster sergeant Nov. 3, 1862. Discharged for disability Oct. 24, 1863. Died July 26, 1911. Interred at St. Augustine National Cemetery. St. Augustine, FL. Plot A, Grave 289.

Surgeon

Harris, James. Residence, Providence. 35. S. Doctor. Commissioned Aug. 18, 1862. Mustered in September 6, 1862. Ordered on special duty as surgeon-in-chief Second Division, Ninth Army Corps, Oct. 19, 1864, and so borne until May 1865. Borne as on special duty as medical director Ninth Army Corps, from May 18, 1865, until June 9, 1865. Brevet lieutenant colonel, U.S. Volunteers for "gallant and meritorious service," March 13, 1865. Mustered out June 9, 1865. Died Aug. 1, 1907. Interred at North Burial Ground, Providence, RI.

Assistant Surgeons

Corey, Charles G. Residence, Providence. 36. M. Doctor. Commissioned and Mustered April 23, 1863. Mustered out July 13, 1865. Died Oct. 19, 1878. Interred at Laurel Hill Cemetery, Fitchburg, MA.

Gaylord, William A. Residence, Providence. 42. M. Doctor. Commissioned Aug. 29, 1862. Mustered in Sept. 6, 1862. Discharged for disability Jan. 2, 1863. Died Oct. 24, 1893. Interred at Oak Grove Cemetery, Pawtucket, RI.

Sprague, Albert G. Residence, Warwick. 36. M. Doctor. Commissioned Aug. 29, 1862. Mustered in September 6, 1862. Borne as absent sick from Nov. 17, 1862 until Jan. 1863. On detached service in hospital at City Point January 1865, and so borne until Mar. 11, 1865. Mustered out June 9, 1865. Died Aug. 1, 1908. Interred at Walnut Grove Cemetery, North Brookfield, MA.

Chaplains

Howard, Harris. Residence, Providence. 42. S. Minister. Commissioned June 4, 1862. Mustered in September 6, 1862. Resigned July 3, 1863.

Sergeant Majors

Allen, Edwin R. Promoted from sergeant Co. A, Feb. 28, 1864. Promoted to first lieutenant Co. A, Oct. 21, 1864.

Fessenden, Samuel. Promoted from private Co. G, June 1, 1863. Promoted to second lieutenant and regimental quartermaster Oct. 20, 1863.

Manchester, Joseph S. Residence, Bristol. 20. S. Clerk. Enlisted Aug. 20, 1862. Mustered in Sept. 6, 1862. Wounded in action, left arm severed, at Fredericksburg, VA, Dec. 13, 1862. Promoted to second lieutenant Co. B, Jan. 7, 1863.

Richter, Henry M. Promoted from sergeant Co. K, Nov. 4, 1864. Mustered out July 13, 1865.

Quartermaster Sergeants

Grafton, Joseph J. D. Promoted from Co. B, Jan. 1863. Mustered out June 9, 1865. Died 1889. Interred at North Burial Ground, Providence, RI.

Stanhope, John R. Jr. Promoted from Co. I, Sept. 7, 1862. Promoted to regimental quartermaster Nov. 3, 1862.

Commissary Sergeant

Clarke, Steadman. Residence, South Kingstown. 39. M. Grocer. Enlisted July 13, 1862. Mustered in Sept. 6, 1862. Mustered out June 9, 1865.

Hospital Steward

Peckham, Stephen F. Residence, North Providence. 23. S. Pharmacist. Enlisted Aug. 15, 1862. Mustered in Sept. 6, 1862. Detached for service at Philadelphia, PA, Jan. 6, 1865, and so borne until May 26, 1865 when mustered out. Died July 11, 1918. Interred at Swan Point Cemetery, Providence, RI.

Principal Musician

Carpenter, James. Promoted from fifer Co. G, Dec. 15, 1864. Mustered out June 9, 1865. Died Sept. 24, 1918. Interred at Oak Dell Cemetery, South Kingstown, RI.

COMPANY A

Captains

Allen, Edward T. Promoted from first lieutenant Co. A, April 1, 1863. Accidentally wounded in leg at Petersburg, VA, June 18, 1864. Discharged for disability Aug. 29, 1864. Died 1926. Interred at Mountain View Cemetery, Oakland, CA.

Leavens, Lewis. Residence, Hopkinton. 39. M. Manufacturer. Commissioned Sept. 4, 1862. Mustered in September 6, 1862. Wounded in action, shot in left thigh, at Fredericksburg, VA, Dec. 13, 1862. Resigned Jan. 12, 1863. Died 1906. Interred at Cypress Hills National Cemetery, Brooklyn, NY.

Peckham, Peleg E. Transferred from Co. B, Oct. 9, 1864. Never served in line as captain of Co. A, remaining on brigade staff as assistant adjutant general. Killed in action at Petersburg, VA, April 2, 1865. Interred at Riverbend Cemetery, Westerly, RI.

First Lieutenants

Allen, Edward T. Promoted from second lieutenant Co. G, Jan. 7, 1863. Promoted to captain Co. A, April 1, 1863.

Allen, Edwin R. Promoted from sergeant major Oct. 26, 1864. Commanding company until mustered out June 9, 1865. Died May 4, 1931. Interred at Pine Grove Cemetery, Hopkinton, RI.

Kenyon, David R. Residence, Richmond. 29. M. Manufacturer. Commissioned Sept. 4, 1862. Mustered in September 6, 1862. Wounded in action, shot in leg, at Fredericksburg, VA, Dec. 13, 1862. Promoted to captain Co. I, Jan. 7, 1863.

Perkins, Benjamin G. Promoted from second lieutenant Co. K, July 1, 1863. Resigned July 20, 1864. Died July 27, 1898. Interred at Moshassuck Cemetery, Central Falls, RI.

Second Lieutenants

Morton, Joseph W. Residence, Hopkinton. 41. M. Teacher. Commissioned Sept. 4, 1862. Mustered in September 6, 1862. Resigned Dec. 4, 1862. Died 1893. Interred at Milton Cemetery, Milton, WI.

Moore, Winthrop A. Promoted from private Co. G, Mar. 1, 1863. Promoted to first lieutenant Co. K, Jan. 9, 1864.

First Sergeant

Barstow, William H. Residence, Providence. 25. M. Artist. Enlisted June 30, 1862. Mustered in Sept. 6, 1862. Wounded in action, shot in stomach at Bethesda Church, VA, June 3, 1864. Mustered out June 9, 1865. Died Sept. 12, 1899. Interred at Pocasset Cemetery, Cranston, RI.

Sergeants

Allen, Edwin R. Promoted from private. Promoted to sergeant major Feb. 28, 1864.

Barber, John N. Promoted from corporal. Mustered out June 9, 1865. Died Nov. 23, 1897. Interred at Rockville Cemetery, Hopkinton, RI.

Carroll, John. Transferred from Co. C, Oct. 21, 1864. Transferred to Veteran Reserve Corps, Jan. 22, 1865.

Cole, Edward C. Resident, Providence. 30. M. Moulder. Enlisted July 21, 1862. Mustered in Sept. 6, 1862. Absent on furlough for fifteen days Mar. 1865. Mustered out June 9, 1865. Died Mar. 5, 1884. Interred at Grace Church Cemetery, Providence, RI.

Flaherty, Michael. Residence, Providence. 21. S. Laborer. Enlisted June 21, 1862. Mustered in Sept. 6, 1862. Wounded in action, shot in leg, at Fredericksburg, VA, Dec. 13, 1862. Sent to hospital and borne as absent sick until Feb. 1863. Killed in action at Bethesda

Church, VA, June 3, 1864. Cenotaph in St. Patrick's Cemetery, Providence, RI.

Lillibridge, Amos A. Promoted from corporal. Killed in action at Spotsylvania Court House, VA, May 18, 1864. Cenotaph in Wood River Cemetery, Richmond, RI.

Peckham, Peleg E. Residence, Charlestown. 27. M. Carpenter. Enlisted Aug. 7, 1862. Mustered in Sept. 6, 1862. Promoted to second lieutenant Co. E, Jan. 12, 1863.

Stoothoff, John B. Promoted from private. Never served as company sergeant, being assigned as regimental color bearer. Mustered out June 9, 1865. Died Nov. 11, 1893. Interred at North Burial Ground, Providence, RI.

Tower, John K. Residence, Hopkinton. 30. M. Farmer. Enlisted Aug. 7, 1862. Mustered in Sept. 6, 1862. Discharged for disability, Aug. 10, 1863.

Corporals

Abbott, George H. Residence, Providence. Enlisted and Mustered as corporal April 4, 1865. Mustered out July 13, 1865.

Barber, John N. Residence, Hopkinton. 29. M. Farmer. Enlisted Aug. 7, 1862. Mustered in Sept. 6, 1862. Promoted to sergeant.

Bowman, George. Residence, Warwick. 30. M. Farmer. Enlisted July 21, 1862. Mustered in Sept. 6, 1862. Mustered out June 9, 1865. Interred at Small Maple Root Cemetery, Coventry, RI.

Gardner, Henry C. Promoted from private. Wounded in action, shot in hand at Spotsylvania Court House, VA, May 12, 1864. Transferred to the Veteran Reserve Corps, Dec. 17, 1864. Died 1923. Interred at Pachaug Cemetery, Griswold, CT.

Lillibridge, Amos A. Residence, Richmond. 18. S. Teacher. Enlisted Aug. 8, 1862. Mustered in Sept. 6, 1862. Promoted to sergeant.

Mallory, William W. Residence, Providence. 29. M. Laborer. Enlisted July 23, 1862. Mustered in Sept. 6, 1862. Discharged for disability at Fort Monroe, VA, April 28, 1863.

Marcoux, Joseph. Residence, Providence. 21. S. Laborer. Enlisted July 10, 1862. Mustered in Sept. 6, 1862. Borne as sick in hospital from Nov. 7, 1862, until Nov. 24, 1862, when he was discharged from the hospital. Mortally wounded in action, shot in throat, at Fredericksburg, VA, Dec. 13, 1862. Died of wounds at Washington, DC, Jan. 7, 1863. Served in Color Guard. Interred at Soldier's Home National Cemetery, Washington, DC. Grave 4600.

Neff, William B. Residence, Glocester. 27. M. Farmer. Enlisted Aug. 5, 1862. Mustered in Sept. 6, 1862. Wounded in action, in left thigh, at Fredericksburg, VA, and sent to hospital, Dec. 13, 1862. Transferred to Veteran Reserve Corps, Nov. 30, 1863. Died 1920. Interred at Acotes Hill Cemetery, Glocester, RI.

Phillips, Reynolds C. Promoted from private. Borne on detached service at New Haven, CT, from Sept. 9, 1863, until June 1865. Mustered out June 15, 1865. Died Mar. 14, 1899. Interred at Wood River Cemetery, Richmond, RI.

Phillips, Oliver J. Promoted from private. Mortally wounded in action, shot in stomach, at Bethesda Church, VA, June 3, 1864. Died of wounds at General Hospital, Washington, DC, July 20, 1864. Interred at Arlington National Cemetery. Section 13. Grave 6489. Cenotaph in Knotty Oak Cemetery, Coventry, RI.

Rathbun, George C. Residence, Richmond. 21. S. Farmer. Enlisted Aug. 11, 1862. Mustered in Sept. 6, 1862. Wounded in action, in left thigh, at Fredericksburg, VA, and sent to hospital, Dec. 13, 1862. Discharged for disability at Portsmouth Grove, RI, June 23, 1863. Interred at Clarke Lot, Richmond Cemetery 33, Richmond, RI.

Wells, Horace. Residence, Hopkinton. 21. S. Farmer. Enlisted Aug. 7, 1862. Mustered in Sept. 6, 1862. Wounded in action, shot in left shoulder, at Fredericksburg, VA, Dec. 13, 1862. Discharged for disability Feb. 1, 1863. Died Mar. 13, 1896. Interred at Pittsfield Cemetery, Pittsfield, MA.

Musicians

Andrews, Albert A. Transferred from Co. C, Feb. 1, 1865. Mustered out June 20, 1865. Died Aug. 28, 1888. Interred at Greenwood Cemetery, Coventry, RI.

Greene, William H. Residence, Providence. 42. M. Mason. Enlisted Aug. 15, 1862, Mustered in Sept. 6, 1862. Promoted from Co. C, Oct. 3, 1862. Died of typhoid at Baltimore, MD, April 21, 1863. Interred at Oakland Cemetery, Cranston, RI.

Hopkins, Charles W. Transferred from Co. D, Feb. 1, 1865. Discharged for disability May 26, 1865. Died June 14, 1910. Interred at Knotty Oak Cemetery, Coventry, RI.

Wagoner

Colwell, William. Residence, Providence. 23. S. Teamster. Enlisted Aug. 5, 1862. Mustered in Sept. 6, 1862. Died Jan. 14, 1863 of typhoid at Falmouth, VA.

Privates

Albro, George B. Residence, Coventry. 19. S. Laborer. Enlisted July 24, 1862. Mustered in Sept. 6, 1862. Wounded in action, in leg, at Fredericksburg, VA, and sent to hospital Dec. 13, 1862. Mustered out June 9, 1865. Died Aug. 14, 1907. Interred at Cottrell Cemetery, Scituate, RI.

Allen, Edwin R. Residence, Hopkinton. 21. S. Clerk. Enlisted Aug. 7, 1862. Mustered in Sept. 6, 1862. Promoted to sergeant.

Arnold, Joseph G. Residence, Hopkinton. 26. M. Machinist. Enlisted Aug. 8, 1862. Mustered in Sept. 6, 1862. On detached service at division headquarters, Jan. 1863. Transferred to Veteran Reserve Corps, Oct. 31, 1863. Died 1916. Interred at Pine Grove Cemetery, Hopkinton, RI.

Austin, Benjamin K. Residence, Hopkinton. 19. S. Farmer. Enlisted Aug. 7, 1862. Mustered in Sept. 6, 1862. Killed in action at Spotsylvania Court House, VA, May 12, 1864. Interred at Fredericksburg National Cemetery. Grave 4168.

Austin, James W. Residence, Hopkinton. 28. M. Wheelwright. Enlisted Aug. 7, 1862. Mustered in Sept. 6, 1862. Teamster in quartermaster's department, April 1865. Clerk in quartermaster's department, April 1865. Mustered out June 9, 1865. Died June 1, 1923. Interred at Knotty Oak Cemetery, Coventry, RI.

Baaden, Theodore. Residence, Providence. 33. M. Confectioner. Enlisted July 30, 1862. Mustered in Sept. 6, 1862. Discharged for disability at Washington, DC, Mar. 19, 1863. Died Feb. 18, 1897. Interred at Oak Woods Cemetery, Chicago, IL.

Barber, Amos P. Jr. Residence, Hopkinton. 18. S. Teamster. Enlisted Aug. 8, 1862. Mustered in Sept. 6, 1862. Temporarily detached to Battery D, 1st RI Light Artillery, Jan. 15, 1863. Returned to 7th RI Vols. Dec. 10, 1864. Mustered out June 9, 1865. Died Feb. 3, 1905. Interred at Pine Grove Cemetery, Hopkinton, RI.

Barber, Charles W. Residence, Charlestown. 17. S. Farmer. Enlisted Aug. 7, 1862. Mustered in Sept. 6, 1862. Discharged for disability at Newport News, VA, Mar. 2, 1863. Died Sept. 10, 1899. Interred at Usquepaugh Cemetery, South Kingstown, RI.

Barber, William A. Residence, Hopkinton. 29. M. Carpenter. Enlisted Aug. 8, 1862. Mustered in Sept. 6, 1862. On extra duty in hospital department Dec. 1862. Mustered out June 9, 1865.

Barlow, Robert S. Residence, Exeter. 34. M. Carpenter. Enlisted July 17, 1862. Mustered in Sept. 6, 1862. Deserted April 16, 1864.

Bentley, William. Residence, North Stonington, CT. 44. M. Weaver. Enlisted Aug. 11, 1862. Mustered in Sept. 6, 1862. On detached service at division headquarters, Nov. 1862, and so borne until Feb. 1863. Killed in a boiler explosion accident at Nicholasville, KY, June 6, 1863.

Bitgood, John F. Residence, Hopkinton. 33. M. Farmer. Enlisted Aug. 11, 1862. Mustered in Sept. 6, 1862. Mustered out June 12, 1865. Died Sept. 27, 1910. Interred at Bitgood Lot, Hopkinton Cemetery 1, Hopkinton, RI.

Briggs, Thomas B. Residence, West Greenwich. 27. M. Machinist. Enlisted Aug. 7, 1862. Mustered in Sept. 6, 1862. Borne as absent on detached services at New Haven, CT, from Sept. 9, 1863, until June 1865. Mustered out June 15, 1865. Died Mar. 26, 1883. Interred at Wood River Cemetery, Richmond, RI.

Brown, George Henry. Residence, Richmond. 20. S. Hatter. Enlisted Aug. 7, 1862. Mustered in Sept. 6, 1862. Wounded in action, shot in leg, at Fredericksburg, VA, Dec. 13, 1862. Wounded in action, shot in leg, at Spotsylvania Court House, VA, and sent to hospital May 12, 1864. Borne as absent sick until April 27, 1865, when he was transferred to the Veteran Reserve Corps. Died 1914. Interred at Pine Grove Cemetery, Hopkinton, RI.

Burdick, John K. Residence, Charlestown. 29. M. Farmer. Enlisted, Aug. 12, 1862. Mustered in Sept. 6, 1862. Discharged for disability at Newport News, VA, Mar. 20, 1863. Interred at Riverbend Cemetery, Westerly, RI.

Burdick, Joseph Weeden. Residence, Hopkinton. 27. M. Farmer. Enlisted Aug. 7, 1862. Mustered in Sept. 6, 1862. Died of Yazoo Fever at Milldale, MS, July 19, 1863. Interred near Milldale, MS. Cenotaph in Rockville Cemetery, Hopkinton, RI.

Burke, Patrick. Residence, Richmond. 24. M. Farmer. Enlisted Aug. 9, 1862. Mustered in Sept. 6, 1862. Wounded in action, shot in leg, at Fredericksburg, VA, Dec. 13, 1862. Wounded in action, shot in arm, at Spotsylvania Court House, VA, May 12, 1864. Absent sick until May 26, 1865, when he was discharged for disability at Augur Hospital, Alexandria, VA.

Cherry, Moses. Residence, Richmond. 24. M. Carder. Enlisted Aug. 12, 1862. Mustered in Sept. 6, 1862. Captured at Fredericksburg, VA, Dec. 13, 1862. Paroled. Transferred to Co. B, June 9, 1865.

Cherry, William. Residence, Cranston. 20. M. Laborer. Enlisted Aug. 9, 1862. Mustered in Sept. 6, 1862. Deserted at Parole Camp, Annapolis, MD, Aug. 30, 1863.

Clark, John Burr. Residence, Richmond. 28. M. Farmer. Enlisted Aug. 8, 1862. Mustered in Sept. 6, 1862. Mortally wounded in action, shot in back, at Fredericksburg, VA, Dec. 13, 1862. Died of wounds at Baltimore, MD, May 10, 1863. Interred at Union Cemetery, North Stonington, CT.

Collins, Gideon Franklin. 24. M. Farmer. Residence, Hopkinton. Enlisted Aug. 8, 1862. Mustered in Sept. 6, 1862. Died of typhoid at Pleasant Valley, MD, Oct. 10, 1862.

Congdon, Oliver H. Residence, North Kingstown. 42. M. Dresser. Enlisted Aug. 9, 1862. Mustered in Sept. 6, 1862. Wounded in action, shot in face, at Jackson, MS, July 13, 1863. Mustered out June 9, 1865. Died July 21, 1892. Interred at Aldrich Cemetery, North Smithfield, RI.

Cundall, Isaac. Residence, Hopkinton. 21. S. Student. Enlisted Aug. 7, 1862. Mustered in Sept. 6, 1862. Mustered out June 9, 1865. Died Jan. 28, 1929. Interred at Evergreen Cemetery, Stonington, CT.

Davis, Martin V. B. Residence, Providence. 21. S. Laborer. Enlisted, July 23, 1862. Mustered in Sept. 6, 1862. Borne as absent sick from Nov. 1862, until Mar. 16, 1863, when he was discharged for disability at Portsmouth Grove, RI.

Donahue, Barney. Residence, Cumberland. 18. S. Glass maker. Enlisted June 27, 1862. Mustered in Sept. 6, 1862. Deserted at Richmond, KY, April 19, 1863.

Donnelly, John. Residence, Providence. 33. M. Weaver. Enlisted June 24, 1862. Mustered in Sept. 6, 1862. Discharged for disability at General Hospital, Fairfax Seminary, Sept. 24, 1864.

Doyle, Michael. Residence, Woonsocket. 25. M. Weaver. Enlisted July 9, 1862. Mustered in Sept. 6, 1862. Mustered out June 9, 1865.

Durfee, William B. Residence, Richmond. 21. S. Farmer. Enlisted Aug. 11, 1862. Mustered in Sept. 6, 1862. Mustered out June 9, 1865. Died 1897. Interred at Oak Grove Cemetery, Hopkinton, RI.

Edwards, Nathan P. Residence, Hopkinton. 21. S. Farmer. Enlisted Aug. 13, 1862, Mustered in Sept. 6, 1862. Wounded in action, shot in back, at Fredericksburg, VA, Dec. 13, 1862. Mustered out June 9, 1865. Died Sept. 22, 1886. Interred at Oak Grove Cemetery, Hopkinton, RI.

Farrell, Edmund. Residence, Burrillville. 21. S. Shoemaker. Enlisted Aug. 4, 1862. Mustered in Sept. 6, 1862. Deserted on march from Wheatland, VA, Nov. 3, 1862.

Flanagan, Bernard. Residence, Providence. 28. S. Laborer. Enlisted July 22, 1862. Mustered in Sept. 6, 1862. Died of Yazoo Fever at Cincinnati, OH, July 25, 1863.

Gardiner, George W. Residence, Hopkinton. 24. M. Laborer Enlisted Aug. 8, 1862. Mustered in Sept. 6, 1862. Died of typhoid at Pleasant Valley, MD, Oct. 18, 1862. Interred at Pine Grove Cemetery, Hopkinton, RI.

Gardner, Henry C. Residence, Hopkinton. 16. S. Farmer. Enlisted Aug. 8, 1862. Mustered in Sept. 6, 1862. Wounded in action, shot in arm, at Fredericksburg, VA, Dec. 13, 1862. Promoted to corporal.

Gardner, John N. Residence, Hopkinton. 21. S. Farmer. Enlisted Aug. 21, 1862. Mustered in Sept. 6, 1862. Discharged for disability at Falmouth, VA, Dec. 7, 1862.

Gates, Hazard R. Residence, Hopkinton. 21. S. Machinist. Enlisted Aug. 11, 1862. Mustered in Sept. 6, 1862. Mustered out June 9, 1865. Died April 3, 1896. Interred at Hoxsie Lot, Hopkinton Cemetery 10, Hopkinton, RI.

Godfrey, Henry H. Residence, Hopkinton. 21. S. Machinist. Enlisted Aug. 8, 1862. Mustered in Sept. 6, 1862. On extra duty in quartermaster's department as teamster Jan. 1863. Discharged for disability at Milldale, MS, July 27, 1863. Died of Yazoo Fever at Hopkinton, RI, September 7, 1863. Interred at Pine Grove Cemetery, Hopkinton, RI.

Gorton, Joel B. Residence, West Greenwich. 21. S. Farmer. Enlisted Aug. 12, 1862. Mustered in Sept. 6, 1862. Wounded in action, in left side, at Fredericksburg, VA, Dec. 13, 1862. Died of Yazoo Fever at Camp Nelson, KY, Sept. 11, 1863. Interred at Camp Nelson National Cemetery. Section D, Grave 1262.

Greene, Charles B. Residence, Hopkinton. 19. S. Farmer. Enlisted Aug. 11, 1862. Mustered in Sept. 6, 1862. Died of typhoid at Frederick, MD Oct. 5, 1862. Interred at First Cemetery, Hopkinton, RI.

Greene, Harris R. Residence, Richmond. 23. M. Spinner. Enlisted Aug. 7, 1862. Mustered in Sept. 6, 1862. Deserted on expiration of furlough June 16, 1863. Interred at Wood River Cemetery, Richmond, RI.

Greene, Jedediah. Residence, Hopkinton. 42. M. Laborer. Enlisted Aug. 14, 1862. Mustered in Sept. 6, 1862. Killed in action at Fredericksburg, VA, Dec. 13, 1862.

Greene, John R. Residence, Hopkinton. 28. M. Farmer. Enlisted Aug. 7, 1862. Mustered in Sept. 6, 1862. Wounded in action, left foot amputated, at Fredericksburg, VA, Dec. 13, 1862. Borne as absent sick until Feb. 6, 1863, when he was discharged for disability at Washington, DC. Interred at Rockville Cemetery, Hopkinton, RI.

Hiscox, John T. Residence, Hopkinton. 21. S. Dresser. Enlisted Aug. 8, 1862. Mustered in Sept. 6, 1862. Wounded in action at Bethesda Church, VA, June 3, 1864. Mustered out June 9, 1865.

Holdridge, Charles H. Residence, Hopkinton. 18. S. Wheelwright. Enlisted Aug. 18, 1862. Mustered in Sept. 6, 1862. Wounded in action, in head, at Fredericksburg, VA, Dec. 13, 1862. Discharged for disability at Newport News, VA, Mar. 2, 1863. Died April 10, 1937. Interred at Oak Grove Cemetery, Hopkinton, RI.

Hudson, Benjamin F. Residence, Providence. 33. M. Clerk. Enlisted Aug. 6, 1862. Mustered in Sept. 6, 1862. In ambulance corps Nov. 1862, and so borne until June 1865. Mustered out June 9, 1865. Interred at Newman-Hunt Cemetery, East Providence, RI.

Hughes, James. Residence, Providence. 45. M. Laborer. Enlisted July 1, 1862. Mustered in Sept. 6, 1862. Drowned in Potomac River, at Aquia Creek, Feb. 9, 1863. Interred at North Burial Ground, Providence, RI.

Jones, John P. Residence, Providence. 32. M. Clerk. Enlisted July 2, 1862. Mustered in Sept. 6, 1862. Mustered out June 9, 1865. Died Mar. 23, 1908. Interred at Jordan Cemetery, Waterford, CT.

Kenyon, Aldrich C. Residence, Hopkinton. 27. M. Farmer. Enlisted Aug. 7, 1862. Mustered in Sept. 6, 1862. Wounded in action at Bethesda Church, VA, June 3, 1864. Mustered out June 9, 1865. Died Nov. 1, 1906. Interred at Pendleton Hill Cemetery, North Stonington, CT.

Kenyon, James G. Residence, Charlestown. 21. S. Farmer. Enlisted Aug. 7, 1862. Mustered in Sept. 6, 1862. Killed in action at Petersburg, VA, June 19, 1864. Interred at Poplar Grove National Cemetery. Grave 3153.

Kenyon, Joseph J. Residence, Hopkinton. 22. S. Farmer. Enlisted Aug. 7, 1862. Mustered in Sept. 6, 1862. Died of typhoid at Falmouth, VA, Nov. 24, 1862. Interred at Pine Grove Cemetery, Hopkinton, RI.

Kenyon, Thomas R. Residence, Hopkinton. 18. S. Farmer. Enlisted Aug. 11, 1862. Mustered in Sept. 6, 1862. Died of Yazoo Fever on board Steamer *David Tatum* on Mississippi River, Aug. 9, 1863.

Langworthy, George W. Residence, Hopkinton. 18. S. Farmer. Enlisted Aug. 11, 1862. Mustered in Sept. 6, 1862. Wounded in action at Petersburg, VA, April 2, 1865. Mustered out June 9, 1865.

Langworthy, Lucius C. Residence, Hopkinton. 23. S. Accountant. Enlisted Aug. 8, 1862. Mustered in Sept. 6, 1862. Discharged for disability Dec. 15, 1862. Died of typhoid at Hopkinton, RI, Jan. 24, 1863. Interred at Pine Grove Cemetery, Hopkinton, RI.

Larkin, Edward. Residence, Richmond. 24. M. Farmer. Enlisted Aug. 8, 1862. Mustered in Sept. 6, 1862. Wounded in action, shot in left knee, at Fredericksburg, VA, Dec. 13, 1862. Mustered out June 9, 1865. Died Dec. 30, 1906. Interred at Wood River Cemetery, Richmond, RI.

Lewis, John D. Residence, Hopkinton. 24. M. Farmer. Enlisted Aug. 16, 1862. Mustered in Sept. 6, 1862. Died of typhoid at Falmouth, VA, Dec. 25, 1862. Interred at Wood River Cemetery, Richmond, RI.

Lewis, George H. Residence, Hopkinton. 20. S. Machinist. Enlisted Aug. 16, 1862. Mustered in Sept. 6, 1862. Mustered out June 9, 1865. Died 1922. Interred at Pine Grove Cemetery, Hopkinton, RI.

McDonough, John. Residence, Burrillville. 23. S. Spinner. Enlisted Aug. 18, 1862. Mustered in Sept 6, 1862. Accidentally shot in the side and sent to hospital June 14, 1864. Transferred to Veteran Reserve Corps, Jan. 10, 1865.

Miner, Edward. Residence, Providence. 21. S. Laborer. Enlisted Aug. 7, 1862. Mustered in Sept. 6, 1862. Deserted June 8, 1863.

Mooney, Patrick J. Residence, Smithfield. 21. S. Laborer. Enlisted Aug. 4, 1862. Wounded in action, shot in wrist, at Spotsylvania Court House, VA, May 18, 1864. Wounded in action at the North Anna River, VA, May 25, 1864. Transferred to Veteran Reserve Corps, Jan. 28, 1865.

Mulvey, Michael. Residence, Cranston. 30. M. Laborer. Enlisted July 18, 1862. Mustered in Sept. 6, 1862. Transferred to Co. C, Oct. 3, 1862.

Nye, James A. Residence, Richmond. 31. M. Carpenter. Enlisted Aug. 7, 1862. Mustered in Sept. 6, 1862. Transferred to the Veteran Reserve Corps, Mar. 11, 1864. Interred at Wood River Cemetery, Richmond, RI.

Palmer, Elisha M. Residence, Hopkinton. 19. S. Farmer. Enlisted Aug. 8, 1862. Mustered in Sept. 6, 1862. Mustered out June 9, 1865. Died July 17, 1909. Interred at First Cemetery, Hopkinton, RI.

Palmer, Henry C. Residence, Hopkinton. 19. S. Farmer. Enlisted Aug. 11, 1862. Mustered in Sept. 6, 1862. Transferred to the Veteran Reserve Corps, Sept. 30, 1863. Died July 27, 1926. Interred at First Cemetery, Hopkinton, RI.

Perry, Albert P. Residence, Hopkinton. 21. M. Spinner. Enlisted Aug. 7, 1862. Mustered in Sept. 6, 1862. Mustered out June 9, 1865. Died Aug. 20, 1920. Interred at Pine Grove Cemetery, Hopkinton, RI.

Phillips, Reynolds C. Residence, Hopkinton. 30. M. Farmer. Enlisted Aug. 7, 1862. Mustered in Sept. 6, 1862. Promoted to corporal.

Phillips, Oliver J. Residence, Coventry. 30. M. Farmer. Enlisted July 18, 1862. Mustered in Sept. 6, 1862. Promoted to corporal.

Richmond, Albert G. Residence, Providence. 23. S. Farmer. Enlisted Feb. 26, 1864. Mustered in Mar. 2, 1864. Mustered out July 13, 1865.

Saunders, Isaac N. Residence, Hopkinton. 21. M. Weaver. Enlisted Aug. 7, 1862. Mustered in Sept. 6, 1862. Killed in action at Spotsylvania Court House, VA, May 12, 1864. Interred at Fredericksburg National Cemetery. Grave 4167. Cenotaph in Rockville Cemetery, Hopkinton, RI.

Slocum, Horace. Residence, Richmond. 20. S. Spinner. Enlisted Aug. 7, 1862. Mustered in Sept. 6, 1862. Wounded in action, shot in hand, at Fredericksburg, VA, Dec. 13, 1862. Wounded in action, shot in back, at the North Anna River, VA, May 24, 1864. Mustered out June 9, 1865. Died Feb. 9, 1917. Interred at Pine Grove Cemetery, Hopkinton, RI.

Stoothoff, John B. Residence, Charlestown. 40. M. Carpenter. Enlisted Sept. 5, 1862. Mustered in Sept. 6, 1862. Promoted to sergeant.

Sunderland, William A. Residence, Hopkinton. 18. S. Operative. Enlisted Aug. 8, 1862. Mustered in Sept. 6, 1862. Mustered out June 9, 1865. Died Jan. 11, 1869. Interred at Mathewson Lot, Coventry Cemetery 68, Coventry, RI.

Taber, Edward S. Residence, Richmond. 27. M. Laborer. Enlisted Aug. 7, 1862. Mustered in Sept. 6, 1862. Wounded in action at Fredericksburg, VA, Dec. 13, 1862 and sent to Washington. Discharged for disability at Baltimore, MD, April 16, 1863. Died Aug. 17, 1904. Interred at Wood River Cemetery, Richmond, RI.

Tallman, Esek B. Residence, Providence. Enlisted and Mustered Mar. 22, 1865. Mustered out July 13, 1865. Died July 2, 1901. Interred at Grace Church Cemetery, Providence, RI.

Thomas, George Arnold. Residence, Hopkinton. 28. M. Blacksmith. Enlisted Aug. 7, 1862. Mustered in Sept. 6, 1862. Died of typhoid at Baltimore, MD, April 14, 1863. Interred at Riverbend Cemetery, Westerly, RI.

Tourgee, Samuel W. Residence, Warwick. 22. S. Laborer. Enlisted July 28, 1862. Mustered in Sept. 6, 1862. Wounded in action, lost finger, at Fredericksburg, VA, Dec. 13, 1862. Wagoner in quartermaster's department, Jan. 1865, and so borne until June 1865. Mustered out June 9, 1865. Died Jan. 13, 1890. Interred at Rogers Lot, Warwick Cemetery 61, Warwick, RI.

Vincent, Charles G. Residence, Hopkinton. 21. S. Teacher. Enlisted Aug. 13, 1862. Mustered in Sept. 6, 1862. Deserted at Pleasant Valley, MD, Oct. 1862.

Weaver, Leander S. Residence, Richmond. 24. S. Farmer. Enlisted Aug. 21, 1862. Mustered in Sept. 6, 1862. Borne as absent sick at Frederick City Hospital, MD, from Oct. 4, 1862, until Jan. 1863. Discharged for disability Jan. 1, 1863.

Weeden, Richard W. Residence, Providence. 18. S. Laborer. Enlisted Aug. 7, 1862. Mustered in Sept. 6, 1862. Wounded at Fredericksburg, VA, and sent to General Hospital, Washington, Dec. 13, 1862. Discharged for disability at Portsmouth Grove, RI, Mar. 16, 1864.

Wells, George C. Residence, Hopkinton. 18. S. Farmer. Enlisted Aug. 11, 1862. Mustered in Sept. 6, 1862. Wounded in action at Fredericksburg, VA, and sent to General Hospital, Dec. 13, 1862. Discharged for disability Jan. 12, 1863.

Whitman, Stephen M. Residence, Hopkinton. 37. M. Soap maker. Enlisted Aug. 8, 1862. Mustered in Sept. 6, 1862. Wounded in action at Fredericksburg, VA, Dec. 13, 1862. Mustered out June 9, 1865.

Wilson, Henry. Residence, Providence. 39. S. Engraver. Enlisted June 27, 1862. Mustered in Sept. 6, 1862. Mustered out June 9, 1865.

Worden, Charles H. Residence, Hopkinton. 18. S. Farmer. Enlisted Aug. 11, 1862. Mustered in Sept. 6, 1862. Detached to Battery D, 1st Rhode Island Light Artillery, Jan. 15, 1863. Died of typhoid at Hampton General Hospital, VA, Feb. 18, 1863. Interred at Hampton National Cemetery. Hampton, VA, Section D, Grave 3292.

Wright, Pardon T. Residence, Richmond. 23. S. Machinist. Enlisted Aug. 12, 1862. Mustered in Sept. 6, 1862. Borne as absent sick at Pleasant Valley, MD, from Oct. 27, 1862, until Feb.

1863. Wounded in action, shot in head, at Petersburg, VA, July 4, 1864. Mustered out June 9, 1865.

COMPANY B

Captains

Peckham, Peleg E. Promoted from first lieutenant Co. E, July 25, 1864. Brevet major for gallantry at Spotsylvania Court House, VA, May 18, 1864. Detached as assistant adjutant general First Brigade, Second Division, Ninth Corps, June 15, 1864. Transferred to Co. A, Oct. 21, 1864.

Winn, Theodore. Residence, Providence. M. 48. Store trimmer. Commissioned Sept. 4, 1862. Mustered in September 6, 1862. Wounded in action, in shoulder, at Fredericksburg, VA, Dec. 13, 1862. Resigned May 18, 1864. Died July 19, 1890. Interred at Arlington National Cemetery. Section 13, Grave 5501.

First Lieutenants

Cole, Darius I. Promoted from first sergeant Co. B, June 25, 1863. Killed in action at Spotsylvania Court House, VA, May 13, 1864. Interred at Fredericksburg National Cemetery. Grave 952.

Chappell, Winfield S. Promoted from sergeant Co. G, Oct. 26, 1864. Transferred to Co. C, Mar. 1865.

Hill, William. Residence, North Providence. M. 24. Carpenter. Commissioned Sept. 4, 1862. Mustered in September 6, 1862. Resigned Oct. 26, 1862. Died Nov. 6, 1900. Interred at Pocasset Cemetery, Cranston, RI.

Manchester, Joseph S. Promoted from second lieutenant Co. B, Mar. 1, 1863. Commissioned captain and commissary of subsistence, United States Volunteers, and transferred to the General Staff, June 25, 1863. Died May 4, 1872. Interred at Juniper Hill Cemetery, Bristol, RI.

Weigand, Frederick. Transferred from Co. G, July 31, 1864. Discharged for disability Sept. 21, 1864. Died May 18, 1900.

Interred at Lexington National Cemetery. Lexington, KY. Grave 1022.

Second Lieutenants

Manchester, Joseph S. Promoted from sergeant major Jan. 7, 1863. Promote to first lieutenant Co. B, Mar. 1, 1863.

Stone, George N. Residence, Providence. 22. S. Merchant. Commissioned Sept. 4, 1862. Mustered in September 6, 1862. On detached service as acting commissary of the First Brigade, Second Division, Ninth Corps, Dec. 1862. Promoted to first lieutenant Co. F, Jan. 18, 1863.

Webb, William W. Promoted from private Co. D, Mar. 1, 1863. Transferred to Co. K, Feb. 1, 1865.

First Sergeants

Bezeley, Jeremiah P. Promoted from sergeant June 25, 1863. Wounded in action, in head, June 8, 1864 at Cold Harbor, VA. Transferred to Co. H, Feb. 1, 1865.

Cole, Darius I. Residence, Providence. 25. M. Machinist. Enlisted July 21, 1862. Mustered in Sept. 6, 1862. Promoted to first lieutenant Co. B, June 25, 1863.

Sergeants

Bezeley, Jeremiah P. Residence, Coventry. 26. M. Painter. Enlisted July 14, 1862. Mustered in Sept. 6, 1862. Promoted to first sergeant June 25, 1863.

Fiske, Alfred. Promoted from private. Wounded in action, shot in hip, at Bethesda Church, VA, June 3, 1864. Transferred to Co. K, Feb. 1, 1865.

Follansbee, Nathan G. Promoted from private. Wounded in action at the Wilderness, VA, May 5, 1864. Transferred to Co. C, Feb. 1, 1865.

Harris, Orrin. Residence, Providence. 53. M. Clerk. Enlisted July 25, 1862. Mustered in Sept. 6, 1862. Discharged for disability Aug. 4, 1864. Died Jan. 1, 1879. Interred at Acotes Hill Cemetery, Glocester, RI.

Gonsolve, Franklin. Residence, Providence. 29. M. Gunsmith. Enlisted July 2, 1862. Mustered in Sept. 6, 1862. Wounded in action, shot in leg, at the Crater, Petersburg, VA, July 30, 1864. Transferred to Co. C, Feb. 1, 1865.

Nottage, John S. Residence, Cranston. 36. M. Carpenter. Enlisted July 5, 1862. Mustered in Sept. 6, 1862. Wounded in action, shot in head, at Spotsylvania Court House, VA, and sent to hospital May 13, 1864. Transferred to Co. K, Feb. 1, 1865.

Corporals

Bennett, Thomas B. Residence, Providence. 47. M. Farmer. Enlisted July 12, 1862. Mustered in Sept. 6, 1862. Transferred to Co. C, Feb. 1, 1865.

Bishop, Charles H. Residence, Providence. 22. S. Seaman. Enlisted July 28, 1862. Mustered in Sept. 6, 1862. Killed in action at Fredericksburg, VA, Dec. 13, 1862. Served in Color Guard.

Bowen, James A. Residence, Barrington. 22. S. Laborer. Enlisted July 4, 1862. Mustered in Sept. 6, 1862. Discharged for disability at Falmouth, VA, Dec. 9 1862. Died Sept. 15, 1884. Interred at North Burial Ground, Bristol, RI.

Bridgehouse, Timothy. Residence, Providence. 38. M. Fisherman. Enlisted July 29, 1862. Mustered in Sept. 6, 1862. Died of typhoid at General Hospital, Camp Dennison, OH, Sept. 14, 1863. Served in Color Guard. Interred at Spring Grove National Cemetery. Grave 1000.

Dennis, Charles E. Residence, Providence. 17. S. Farmer. Enlisted July 8, 1862. Mustered in Sept. 6, 1862. Transferred to Co. C, Feb. 1, 1865. Served in Color Guard.

Higgins, Thomas J. Residence, Providence. 34. S. Oysterman. Enlisted Aug. 6, 1862. Mustered in Sept. 6, 1862. Transferred to Co. C, Feb. 1, 1865.

Swarts, George A. Residence, Providence. 32. S. Sailor. Enlisted July 2, 1862. Mustered in Sept. 6, 1862. Wounded in action at Fredericksburg, VA, Dec. 13, 1862. Discharged for disability at Providence May 10, 1863. Died April 19, 1867. Interred at Swan Point Cemetery, Providence, RI.

Weeden, James. Promoted from private. Transferred to Veteran Reserve Corps, Sept. 17, 1863. Died Mar. 6, 1914. Interred at Swan Point Cemetery, Providence, RI.

Whitcomb, Lyman. Promoted from private. Killed in action at Spotsylvania Court House, VA, May 17, 1864. Interred at Fredericksburg National Cemetery. Grave 282.

Musician

Sheldon, Nehemiah. Residence, North Providence. 37. M. Carpenter. Enlisted Aug. 1862. Mustered in Sept. 6, 1862. Transferred to Co. C, Feb. 1, 1865.

Privates

Billington, Daniel R. Residence, North Providence. 18. S. Laborer. Enlisted July 28, 1862. Mustered in Sept. 6, 1862. Transferred to Co. G, Nov. 20, 1862.

Bingham, Joseph A. Residence, Andover, CT. 21. S. Farmer. Enlisted June 17, 1862. Mustered in Sept. 6, 1862. Discharged for disability from hospital at Philadelphia, PA, Nov. 12, 1862. Died July 19, 1901. Interred at New Andover Cemetery, Andover, CT.

Bitgood, Joseph A. Residence, Providence. 22. M. Spinner. Enlisted July 25, 1862. Mustered in Sept. 6, 1862. Sick in Ninth Army Corps Hospital Dec. 1862. Died in hospital of typhoid at Washington, Jan. 4, 1863. Interred at Soldier's Home National Cemetery, Washington, DC. Grave 5465.

Brennan, Michael. Residence, Johnston. 25. S. Carder. Enlisted July 5, 1862. Mustered in Sept. 6, 1862. Absent on furlough for fifteen days Jan. 1865. Transferred to Co. C, Feb. 1, 1865.

Brickley, James. Residence, Woonsocket. 19. S. Fisherman. Enlisted July 1, 1862. Mustered in Sept. 6, 1862. Killed in action at Fredericksburg, VA, Dec. 13, 1862.

Case, William S. Residence, North Kingstown. 37. M. Laborer. Enlisted July 22, 1862. Mustered in Sept. 6, 1862. Deserted at Falmouth, VA, Jan. 21, 1863. Arrested and placed in confinement Jan. 1864. Deserted while in confinement at headquarters Second Division, Ninth Army Corps, awaiting sentence of general court-martial for desertion May 17, 1865.

Caswell, Alfred A. Residence, Scituate. 23. S. Laborer. Enlisted July 7, 1862. Mustered in Sept. 6, 1862. Died of dysentery in Regimental Hospital at Lexington, KY, Sept. 22, 1863. Interred at Lexington National Cemetery. Grave 467.

Carpenter, Richard. Residence, Johnston. 22. S. Laborer. Enlisted Aug. 5, 1862. Mustered in Sept. 6, 1862. Wounded in action, shot in leg, at Petersburg, VA, June 21, 1864. Transferred to Co. C, Feb. 1, 1865.

Collins, Edward F. Residence, Providence. 18. S. Spinner. Enlisted July 18, 1862. Mustered in Sept. 6, 1862. Transferred to Co. C, Feb. 1, 1865.

Collins, James D. Residence, Providence. 35. M. Engineer. Enlisted July 19, 1862. Mustered in Sept. 6, 1862. Wounded in action at Fredericksburg, VA, Dec. 13, 1862. Wounded in action, shot in leg, at Spotsylvania Court House, VA, May 18, 1864, and sent to hospital. Deserted at Filbert Street Hospital, Philadelphia, July 3, 1864.

Collins, Patrick. Residence, Providence. 18. S. Laborer. Enlisted July 28, 1862. Mustered in Sept. 6, 1862. Wounded in action at Fredericksburg, VA, and sent to hospital, Dec. 13, 1862.

Discharged for disability at Baltimore, MD, April 10, 1863. Died April 16, 1905. Interred at Day Family Cemetery, Vienna, VA.

Cornell, Lewis E. Residence, Providence. 18. S. Farmer. Enlisted July 19, 1862. Mustered in Sept. 6, 1862. Transferred to Co. C, Feb. 1, 1865.

Courtney, William. Residence, Providence. 45. M. Weaver. Enlisted July 28, 1862. Mustered in Sept. 6, 1862. Transferred to Co. C, Feb. 1, 1865.

Cox, William. Residence, Providence. 35. M. Gardiner. Enlisted June 27, 1862. Mustered in Sept. 6, 1862. Killed in action at Fredericksburg, VA, Dec. 13, 1862.

Coyle, Joseph. Residence, Smithfield. 18. S. Farmer. Enlisted Aug. 22, 1862. Mustered in Sept. 6, 1862. Discharged for disability at Pleasant Valley, MD, Oct. 25, 1862. Died July 31, 1888. Interred at Togus National Cemetery. Grave 672.

Crane, Thomas. Residence, Woonsocket. 17. S. Spinner. Enlisted July 1, 1862. Mustered in Sept. 6, 1862. Died of dysentery in General Hospital at Lexington, KY, Nov. 7, 1863. Interred at Lexington National Cemetery. Grave 471.

Creyton, Samuel. Residence, Woonsocket. 41. M. Spinner. Enlisted July 11, 1862. Mustered in Sept. 6, 1862. Discharged for disability at Newport News, VA, Mar. 5, 1863. Died 1876. Interred at North Burial Ground, Providence, RI.

Danforth, Ozias C. Residence, Providence. 47. M. Jeweler. Enlisted June 27, 1862. Mustered in Sept. 6, 1862. Discharged for disability at Newport News, VA, Mar. 4, 1863. Died December 27, 1891. Interred at North Burial Ground, Bristol, RI.

Daniels, Herbert. Residence, Woonsocket. 25. S. Engineer. Enlisted June 30, 1862. Mustered in Sept. 6, 1862. Transferred to Co. C, Feb. 1, 1865.

Dean, Isaac N. Residence, North Providence. 34. S. Laborer. Enlisted Aug. 15, 1862. Mustered in Sept. 6, 1862. Transferred to Co. C, Feb. 1, 1865.

Dugan, Hugh. Residence, Providence. 22. S. Laborer. Enlisted Aug. 22, 1862. Mustered in Sept. 6, 1862. Transferred to Co. C, Feb. 1, 1865.

Dunn, Thomas. Residence, Woonsocket. 17. S. Glass maker. Enlisted June 27, 1862. Mustered in Sept. 6, 1862. Deserted at Covington, KY, Mar. 30, 1863.

Edwards, Robert. Residence, Providence. 35. M. Printer. Enlisted July 25, 1862. Mustered in Sept. 6, 1862. Deserted at Columbus, OH, Mar. 29, 1863.

Farnum, Edwin A. Residence, Glocester. 23. S. Farmer. Enlisted Aug. 6, 1862. Mustered in Sept. 6, 1862. Wounded in action, shot in head, at Spotsylvania Court House, VA, May 13, 1864. Transferred to Co. C, Feb. 1, 1865.

Ferry, James. Residence, Woonsocket. 18. S. Farmer. Enlisted July 5, 1862. Mustered in Sept. 6, 1862. Died of typhoid at Hampton, VA, Mar. 22, 1863.

Fiske, Alfred. Residence, North Providence. 30. S. Merchant. Enlisted Aug. 13, 1862. Mustered in Sept. 6, 1862. Promoted to sergeant.

Fleming, Thomas. Residence, Providence. 19. S. Sailor. Enlisted July 7, 1862. Mustered in Sept. 6, 1862. Wounded in action, shot in finger, at Bethesda Church, VA, June 3, 1864. Transferred to Co. C, Feb. 1, 1865.

Foley, Dennis. Residence, Providence. 30. M. Laborer. Enlisted Aug. 2, 1862. Mustered in Sept. 6, 1862. Wounded in action at Fredericksburg, VA, Dec. 13, 1862. Wounded in action, shot in hand, at Jackson, MS, July 13, 1863. Transferred to Co. C, Feb. 1, 1865.

Follansbee, Nathan G. Residence, Providence. 41. M. Mason. Enlisted Aug. 11, 1862. Mustered in Sept. 6, 1862. Promoted to sergeant.

Fuller, Delmont E. Residence, North Providence. 18. S. Jeweler. Enlisted July 18, 1862. Mustered in Sept. 6, 1862. Discharged for disability at Camp Dennison, OH, Oct. 16, 1863.

Grafton, Joseph J. D. Residence, Cranston. 42. M. Carpenter. Enlisted July 18, 1862. Mustered in Sept. 6, 1862. Transferred to non-commissioned staff as quartermaster sergeant Jan. 1863.

Goodwin, Horton. Residence, Providence. 26. S. Sailor. Enlisted July 21, 1862. Mustered in Sept. 6, 1862. Deserted in the face of the enemy at Fredericksburg, VA, Dec. 13, 1862.

Hayward, Joshua S. Residence, Boston, MA. 59. M. Printer. Enlisted July 28, 1862. Mustered in Sept. 6, 1862. Discharged for disability at Pleasant Valley, MD, Oct. 25, 1862.

Jillson, Stephen C. Residence, Smithfield. 36. M. Farmer. Enlisted July 26, 1862. Mustered in Sept. 6, 1862. Transferred to Veteran Reserve Corps. Nov. 11, 1863.

Johnson, John R. Residence, Providence. 21. S. Laborer. Enlisted July 21, 1862. Mustered in Sept. 6, 1862. On extra duty in quartermaster's department as teamster Jan. 1863. Transferred to Veteran Reserve Corps, Dec. 12, 1863. Interred at Prospect Cemetery, Uxbridge, MA.

Kay, James. Residence, Providence. Assigned to Co. B, 7th Rhode Island Vols. Feb. 8, 1864, to serve out time for his enlistment having deserted at the 12th Rhode Island Vols. in 1862. Died of dysentery in Harewood Hospital, Washington, DC, Sept. 28, 1864.

Kelly, John. Residence, Providence. 23. S. Operative. Enlisted July 9, 1862. Mustered in Sept. 6, 1862. Deserted at Covington, KY, Mar. 30, 1863.

Lamby, Peter. Residence, Smithfield. 20. S. Laborer. Enlisted July 31, 1862. Mustered in Sept. 6, 1862. Wounded in action, shot in arm, at Spotsylvania Court House, VA, May 13, 1864. Transferred to Co. C, Feb. 1, 1865.

Lanahan, Michael. Residence, Providence. 19. S. Laborer. Enlisted July 8, 1862. Mustered in Sept. 6, 1862. Deserted April 8, 1864.

Lane, John P. Residence, Boston, MA. 23. S. Grocer. Enlisted July 16, 1862. Mustered in Sept. 6, 1862. Wounded in action at Fredericksburg, VA, Dec. 13, 1862. Discharged for disability April 7, 1863, at Baltimore, MD.

Laugherty, John. Residence, Providence. 37. M. Spinner. Enlisted July 7, 1862. Mustered in Sept. 6, 1862. On detached service in ambulance corps Nov. 1862, and so borne until June 1865. Transferred to Co. C, Feb. 1, 1865.

Lynch, John. Residence, Providence. 30. M. Weaver. Enlisted July 18, 1862. Mustered in Sept. 6, 1862. Mortally wounded in action at Fredericksburg, VA, Dec. 13, 1862. Died of wounds at Trinity General Hospital, Dec. 25, 1862. Interred at Soldier's Home National Cemetery, Washington, DC. Grave 3813.

McCafferey, Patrick. Residence, Providence. 40. M. Farmer. Enlisted July 1, 1862. Mustered in Sept. 6, 1862. Transferred to Co. C, Feb. 1, 1865.

McCready, Daniel. Residence, Providence. 39. M. Printer. Enlisted July 16, 1862. Mustered in Sept. 6, 1862. Wounded in action at Poplar Spring Church, VA, Sept. 30, 1864. Transferred to Co. C, Feb. 1, 1865.

McElroy, Daniel Residence, Providence. 22. M. Laborer. Enlisted June 25, 1862. Mustered in Sept. 6, 1862. Discharged for disability at Baltimore, MD, May 12, 1863.

McGann, James. Residence, Providence. 17. S. Laborer. Enlisted June 27, 1862. Mustered in Sept. 6, 1862. Deserted at camp at Richmond, KY, April 19, 1863.

McGuinn, James. Residence, Providence. 22. S. Printer. Enlisted July 29, 1862. Mustered in Sept. 6, 1862. Wounded in action at Fredericksburg, VA, Dec. 13, 1862. Borne as absent sick until Jan. 1863. Transferred to Veteran Reserve Corps, Aug. 15, 1863. Interred at St. Patrick's Cemetery, Providence, RI.

McLaughlin, Neil. Residence, Boston, MA. 30. S. Laborer. Enlisted July 16, 1862. Mustered in Sept. 6, 1862. Transferred to Co. C, Feb. 1, 1865.

McLeroy, William. Residence, Providence. 50. M. Laborer. Enlisted July 8, 1862. Mustered in Sept. 6, 1862. Discharged for disability at Baltimore, MD, June 20, 1863.

Marchant, Elisha A. Residence, Providence. 41. M. Printer. Enlisted July 18, 1862. Mustered in Sept. 6, 1862. Wounded in action, in hand at Spotsylvania Court House, VA, May 18, 1864. Transferred to Co. C, Feb. 1, 1865.

Matthewson, Cornelius. Residence, North Providence. 37. M. Mason. Enlisted Aug. 31, 1862. Mustered in Sept. 6, 1862. Detached for service in ambulance corps from Nov. 1862, until Feb. 1863. Transferred to Co. C, Feb. 1, 1865.

Mitchell, Gideon S. Residence, Providence. 18. S. Farmer. Enlisted Aug. 2, 1862. Mustered in Sept. 6, 1862. Wounded in action, in side, at Fredericksburg, VA, Dec. 13, 1862. Sent to Washington Dec. 20, 1862. Borne as absent sick until Feb. 1863. Discharged for disability at Portsmouth Grove, RI, Nov. 19, 1863.

Moran, Patrick. Residence, Providence. 31. M. Laborer. Enlisted July 10, 1862. Mustered in Sept. 6, 1862. Transferred to Co. C, Feb. 1, 1865.

Mulvey, Thomas. Residence, Providence. 40. M. Laborer. Enlisted and Mustered Aug. 10, 1864. Killed in action at Poplar Spring Church, VA, Sept. 30, 1864.

Neil, Horace. Residence, Providence. 22. S. Farmer. Enlisted Aug. 19, 1862. Mustered in Sept. 6, 1862. Discharged for disability Jan. 24, 1864 at New York, NY.

Norton, Sylvanus E. Residence, Burrillville. Enlisted Aug. 4, 1862. Mustered in Sept. 6, 1862. On detached service as orderly at brigade headquarters Jan. 1863. Transferred to Veteran Reserve Corps, Mar. 1864. Died Nov. 11, 1895. Interred at East Union Presbyterian Cemetery, Dilltown, PA.

Nye, Byron D. Residence, Richmond. 18. S. Farmer. Enlisted July 4, 1862. Mustered in Sept. 6, 1862. Wounded in action, shot in hand, at Jackson, MS, July 13, 1863. Transferred to Co. C, Feb. 1, 1865.

Pearce, Holden. Residence, North Providence. 46. M. Carpenter. Enlisted Aug. 15, 1862. Mustered in Sept. 6, 1862. Wounded in action, in thigh and head, at Fredericksburg, VA, Dec. 13, 1862. Wounded in action, shot in shoulder, at Spotsylvania Court House, VA, May 12, 1864. Discharged for disability at Portsmouth Grove, RI, Jan. 3, 1865. Died Feb. 2, 1889. Interred at Swan Point Cemetery, Providence, RI.

Quinlan, Thomas. Residence, Providence. 20. S. Laborer. Enlisted July 28, 1862. Mustered in Sept. 6, 1862. Wounded in action, shot through thighs, at Poplar Spring Church, VA, Oct. 1, 1864. Transferred to Co. C, Feb. 1, 1865.

Riley, John. Residence, Providence. 18. S. Spinner. Enlisted July 18, 1862. Mustered in Sept. 6, 1862. Transferred to Co. C, Feb. 1, 1865.

Rowan, Thomas. Residence, Providence. 42. M. Carrier. Enlisted July 16, 1862. Mustered in Sept. 6, 1862. Died of dysentery in General Hospital, Covington, KY, Aug. 13, 1863.

Robley, George W. Residence, Providence. 43. M. Ferryman. Enlisted Aug. 4, 1862. Mustered in Sept. 6, 1862. Discharged for disability at Fort Columbus, NY Harbor, Mar. 18, 1863. Died of typhoid contracted in the service at Providence, RI, June 19, 1863.

Sanford, William. Residence, Providence. 39. M. Harness maker. Enlisted July 2, 1862. Mustered in Sept. 6, 1862. Wounded in action, in head, at Fredericksburg, VA, 13, 1862. Deserted at hospital at Camp Dennison, OH, July 24, 1863.

Sheridan, John. Residence, Providence. 21. S. Weaver. Enlisted July 26, 1862. Mustered in Sept. 6, 1862. Transferred to Co. C, Feb. 1, 1865.

Smith, George A. Residence, Providence. 32. M. Carpenter. Enlisted Aug. 9, 1862. Mustered in Sept. 6, 1862. Deserted on expiration of furlough April 9, 1863.

Spencer, William H. Residence, Providence. 22. M. Farmer. Enlisted July 7, 1862. Mustered in Sept. 6, 1862. Died of Yazoo Fever onboard *David Tatum*, Aug. 11, 1863. Interred at Napoleonville, AR. Cenotaph in Moses Spencer Lot, East Greenwich Cemetery 28, East Greenwich, RI.

Stafford, Charles W. Residence, Glocester. 20. S. Farmer. Enlisted Aug. 5, 1862. Mustered in Sept. 6, 1862. Transferred to Veteran Reserve Corps, Aug. 15, 1863. Died 1919. Interred at Thomas Stafford Lot, Warwick Cemetery 53, Warwick, RI.

Steele, Daniel. Residence, Glocester. 18. S. Farmer. Enlisted Aug. 6, 1862. Mustered in Sept. 6, 1862. Deserted at Cincinnati, OH, Mar. 30, 1863.

Steere, John F. Residence, Smithfield. 27. M. Farmer. Enlisted Aug. 9, 1862. Mustered in Sept. 6, 1862. Died of Yazoo Fever in Regimental Hospital at Lexington, KY, Oct. 1, 1863. Interred at Lexington National Cemetery. Grave 465.

Thomas, David E. Residence, North Providence. 28. M. Machinist. Enlisted Aug. 13, 1862. Mustered in Sept. 6, 1862. Discharged for disability at Falmouth, VA, Jan. 31, 1863. Died July 11, 1868. Interred at Moshassuck Cemetery, Central Falls, RI.

Weeden, James. Residence, North Providence. 31. M. Seaman. Enlisted Aug. 5, 1862. Mustered in Sept. 6, 1862. Promoted to corporal.

Williams, Cyrus D. Residence, Smithfield. 21. S. Blacksmith. Enlisted Aug. 11, 1862. Mustered in Sept. 6, 1862. Sick at Philadelphia, PA, Dec. 1862. Discharged for disability at Philadelphia, Feb. 9, 1863. Died of disease contracted in the service at Smithfield, RI, Aug. 21, 1865. Interred in Williams Lot, Smithfield Cemetery 17, Smithfield, RI.

Weaver, Stephen. Residence, Providence. 36. M. Ferryman. Enlisted July 28, 1862. Mustered in Sept. 6, 1862. Transferred to Veteran Reserve Corps Oct. 31, 1863. Died Mar. 28, 1888. Interred at Oak Grove Cemetery, Hopkinton, RI.

Whipple, John G. Residence, Providence. 22. S. Farmer. Enlisted Aug. 19, 1862. Mustered in Sept. 6, 1862. Wounded in action, shot in wrist, at Petersburg, VA, July 1, 1864. Discharged for disability May 5, 1865. Died 1910. Interred at Knotty Oak Cemetery, Coventry, RI.

Whitcomb, Lyman. Residence, Barrington. 30. M. Painter. Enlisted Aug. 2, 1862. Mustered in Sept. 6, 1862. Promoted to corporal.

Wright, Harris C. Residence, Burrillville. 33. S. Jeweler. Enlisted June 28, 1862. Mustered in Sept. 6, 1862. Killed in action at Fredericksburg, VA, Dec. 13, 1862. Cenotaph at Locust Grove Cemetery, Providence, RI.

Willard, Nathan F. Residence, Providence. 44. M. Blacksmith. Enlisted Aug. 19, 1862. Mustered in Sept. 6, 1862. Discharged for disability at Washington, DC, Dec. 16, 1862.

COMPANY C

Captains

Church, George E. Residence, Providence. 27. S. Engineer. Commissioned July 26, 1862. Mustered in September 6, 1862. Acting lieutenant colonel of the regiment Dec. 13, 1862. Promoted to lieutenant colonel Jan. 7, 1863.

Potter, James N. Promoted from first lieutenant Co. C, Mar. 1, 1863. Wounded in action at Spotsylvania Court House, VA, May 18, 1864. Detached for service at Concord, NH, Dec. 30, 1864, and so borne until May 25, 1865. Mustered out July 2, 1865. Died of disease contracted in the service Nov. 1, 1869 at Providence, RI. Interred at Island Cemetery, Newport, RI.

First Lieutenants

Chappell, Winfield S. Transferred from Co. B, Feb. 1, 1865. Mustered out June 9, 1865. Died Aug. 29, 1916. Interred at Riverside Cemetery, South Kingstown, RI.

Lincoln, Henry. Promoted from second lieutenant Co. C, Mar. 1, 1863. Resigned April 20, 1864. Died April 6, 1903. Interred at Moshassuck Cemetery, Central Falls, RI.

Remington, Daniel S. Residence, Providence. 25. S. Clerk. Commissioned Dec. 6, 1861. Transferred from Co. A, Fifth RI Heavy Artillery May 7, 1864. Transferred to Co. G, Feb. 1, 1865.

Potter, James N. Promoted from second lieutenant Co. C, Jan. 7, 1863. Promoted to captain Co. C, Mar. 1, 1863.

Second Lieutenants

Lincoln, Henry. Promoted from sergeant Co. I, Jan. 7, 1863. Promoted to first lieutenant Co. C, Mar. 1, 1863.

Merrill, James F. Promoted from sergeant Co. D, Mar. 1, 1863. Promoted to first lieutenant Co. D, July 1, 1863.

Potter, James N. Residence, Providence. 24. S. Clerk. Commissioned Aug. 6, 1862. Mustered in Sept. 6, 1862. Promoted first lieutenant Co. C, Jan. 7, 1863.

First Sergeants

Jenckes, Alonzo L. Promoted from corporal Oct. 30, 1863. Mustered out June 9, 1865.

Westcott, David B. Residence, Johnston. 32. M. Overseer. Enlisted Aug. 5, 1862. Mustered in Sept. 6, 1862. Borne on detached service at division headquarters from Dec. 22, 1862 until Feb. 1863. Died of dysentery at Lexington, KY Oct. 26, 1863. Interred at Lexington National Cemetery. Grave 470.

Sergeants

Batchelder, George T. Residence, North Providence. 26. S. Clerk. Enlisted Aug. 6, 1862. Mustered in Sept. 6, 1862. Wounded in action, in head, at Fredericksburg, VA, Dec. 13, 1862. Wounded in action at Spotsylvania Court House, VA, May 12, 1864. Wounded in action, shot in back, at Spotsylvania Court House, VA, May 18, 1864. Mustered out June 9, 1865. Served in Color Guard. Died 1919. Interred at Acotes Hill Cemetery. Glocester, RI.

Brown, Joseph R. Residence, Coventry. 33. M. Dresser. Enlisted Aug. 5, 1862. Mustered in Sept. 6, 1862. Mustered out June 9, 1865.

Carroll, John. Transferred from Co I, Mar. 10, 1863. Borne as absent sick at Portsmouth Grove, RI, from June 22, 1863, until Jan. 22, 1864, when he was transferred to Veteran Reserve Corps.

Dunbrack, George E. Residence, Warwick. 33. M. Carpenter. Enlisted, Aug. 5, 1862. Mustered in Sept. 6, 1862. Deserted at Lexington, KY, April 1, 1863.

Follansbee, Nathan G. Transferred from Co. B, Feb. 1, 1865. Mustered out June 9, 1865. Died July 20, 1899. Interred at Little Neck Cemetery, East Providence, RI.

Gonsolve, Franklin. Transferred from Co. B, Feb. 1, 1865. Mustered out June 9, 1865. Died Aug. 17, 1895. Interred at Swan Point Cemetery, Providence, RI.

Morse, Henry L. Residence, Coventry. 32. M. Moulder. Enlisted Aug. 5, 1862. Mustered in Sept. 6, 1862. Borne on detached service in ambulance corps from Oct. 26, 1862, until Feb. 1863. Died of dysentery at Annapolis, MD, April 12, 1864. Interred at North Burial Ground, Providence, RI.

Smith, Orlando. Promoted from corporal. Wounded in action, in shoulder, at Bethesda Church, VA, June 3, 1864. Wounded in action, shot in foot, at Petersburg, VA, July 11, 1864. Mustered out June 9, 1865.

Corporals

Aldrich, Moses H. Residence, Burrillville. 34. M. Dresser. Enlisted Aug. 5, 1862. Mustered in Sept. 6, 1862. Wounded in action, shot in leg, at Petersburg, VA, June 29, 1864. Absent sick at Washington, DC, Jan. 1865, and so borne until May 1865. Absent sick at Boston, MA, May 1865. Discharged for disability at Boston, MA, July 12, 1865. Died of disease contracted in the service at Burrillville, RI, Dec. 17, 1865. Interred at Cooper-Mowry Lot, Burrillville Cemetery 46, Burrillville, RI.

Chace, John H. Residence, Providence. 26. M. Carpenter. Enlisted Aug. 9, 1862. Mustered in Sept. 6, 1862. Wounded in action at Fredericksburg, VA, Dec. 13, 1862. Discharged for disability April 1, 1864. Died of dysentery contracted in the service at Providence, RI June 20, 1865. Interred at Swan Point Cemetery, Providence, RI.

Dennis, Charles E. Transferred from Co. B, Feb. 1, 1865. Mustered out June 9, 1865. Died 1932. Interred at Locust Grove Cemetery, Providence, RI.

Gardner, Nelson. Residence, South Kingstown. 37. M. Dresser. Enlisted Aug. 5, 1862. Mustered in Sept. 6, 1862. Discharged for disability at Camp Dennison, OH, Oct. 12, 1863.

Higgins, Thomas J. Transferred from Co. B, Feb. 1, 1865. Promoted to sergeant and transferred to Co. E, Feb. 15, 1865.

Howarth, Abraham H. Residence, Richmond. 26. M. Machinist. Enlisted Aug. 5, 1862. Mustered in Sept. 6, 1862. Mortally wounded in action at Fredericksburg, VA, Dec. 13, 1862. Died of wounds Dec. 19, 1862, at Falmouth, VA. Interred at Fredericksburg National Cemetery. Grave 2344.

Jenckes, Alonzo L. Residence, Smithfield. 18. S. Farmer. Enlisted Aug. 9, 1862. Mustered in Sept. 6, 1862. Wounded in action at Fredericksburg, VA, Dec. 13, 1862. Promoted to first sergeant Oct. 30, 1863.

Potter, Francis W. Promoted from private. Mortally wounded in action at Spotsylvania Court House, VA, May 13, 1864. Died of wounds at Fredericksburg, VA, May 20, 1864. Interred at Fredericksburg National Cemetery. Grave 285.

Smith, Orlando. Residence, Barrington. 27. M. Varnisher. Enlisted Aug. 2, 1862. Mustered in Sept. 6, 1862. Promoted to sergeant.

Willey, Abel W. Residence, Johnston. 36. M. Carpenter. Enlisted Aug. 6, 1862. Mustered in Sept. 6, 1862. Transferred to Veteran Reserve Corps Aug. 6, 1864. Died Feb. 24, 1893. Interred at Greenville Cemetery, Smithfield, RI.

Wilson, Benjamin A. Transferred from Co. G, Feb. 1, 1865. Discharged for disability at West Philadelphia, PA, May 17, 1865. Served in Color Guard. Died Nov. 27, 1904. Interred at Elm Grove Cemetery, North Kingstown, RI.

Musicians

Andrews, Albert A. Residence, Cranston. 26. S. Machinist. Enlisted Aug. 8, 1862. Mustered in Sept. 6, 1862. Transferred to Co. A, Feb. 1, 1865.

Sheldon, Nehemiah. Transferred from Co. B, Feb. 1, 1865. Absent sick at Portsmouth Grove, RI, until June 1865. Mustered out at Providence June 16, 1865.

Winsor, Richard A. Residence, Johnston. 28. M. Farmer. Enlisted Aug. 5, 1862. Mustered in Sept. 6, 1862. Absent sick Jan. 1865, and so borne until May 1865. Mustered out May 29, 1865. Died May 19, 1911. Interred at Samuel Winsor Lot, Johnson Cemetery 47, Johnson, RI.

Wagoner

Coman, William A. Residence, Glocester. 26. M. Farmer. Enlisted Aug. 4, 1862. Mustered in Sept. 6, 1862. Mortally wounded in action at Fredericksburg, VA, Dec. 13, 1862. Died of wounds at Falmouth, VA, Dec. 19, 1862. Interred at Fredericksburg National Cemetery. Grave 4899. Cenotaph in Coman Lot, Glocester Cemetery 48, Glocester, RI.

Privates

Adams, Sabine G. Residence, Glocester. 19. S. Farmer. Enlisted Aug. 7, 1862. Mustered in Sept. 6, 1862. Died of typhoid at Baltimore, MD, Jan. 20, 1863. Interred at Loudon Park National Cemetery. Baltimore, MD. Grave 1687.

Allen, John F. Residence, Glocester. 29. M. Carpenter. Enlisted Aug. 6, 1862. Mustered in Sept. 6, 1862. Wounded in action, shot in hand at Spotsylvania Court House, VA, May 18, 1864. Absent sick at Washington, DC, Jan. 1865, and so borne until April 1865. Transferred to Veteran Reserve Corps April 1, 1865. Died Nov. 23, 1896. Interred at Swan Point Cemetery, Providence, RI.

Arnold, Benjamin F. Residence, Coventry. 18. S. Farmer. Enlisted and Mustered Feb. 15, 1865. Transferred to Co. B, June 9, 1865.

Arnold, Henry A. Residence, Johnston. 27. M. Painter. Enlisted Aug. 6, 1862. Mustered in Sept. 6, 1862. Wounded in action, shot in shoulder, at Petersburg, VA, Aug. 9, 1864. Absent sick at City Point, Jan. 1865. Mustered out June 9, 1865. Died Dec. 16, 1896. Interred at Pocasset Cemetery, Cranston, RI.

Arnold, William C. Residence, Johnston. 32. M. Moulder. Enlisted Aug. 5, 1862. Mustered in Sept. 6, 1862. Absent sick at City Point, VA, Jan. to Mar. 1865. Mustered out June 9, 1865. Died Oct. 24, 1902. Interred at Greenwood Cemetery, Coventry, RI.

Barry, David. Residence, Portsmouth. 26. M. Machinist. Enlisted and Mustered Mar. 2, 1865. Transferred to Co. B, June 9, 1865.

Beebe, Henry W. Residence, Smithfield 18. S. Farmer. Enlisted Aug. 18, 1862. Mustered in Sept. 6, 1862. Deserted Jan. 28, 1863.

Bennett, Thomas B. Transferred from Co. B, Feb. 1, 1865. Mustered out June 9, 1865. Died Oct. 21, 1891. Interred at Spears Cemetery, Foster, RI.

Blackman, James A. Residence, Glocester. 29. S. Farmer. Enlisted Aug. 7, 1862. Mustered in Sept. 6, 1862. Transferred to the Veteran Reserve Corps Oct. 29, 1863.

Brennan, Michael. Transferred from Co. B, Feb. 1, 1865. Mustered out June 9, 1865. Died Aug. 27, 1890. Interred at St. Patrick's Cemetery. East Greenwich, RI.

Brown, John, 1st. Residence, Johnston. 29. S. Farmer. Enlisted Aug. 5, 1862. Mustered in Sept. 6, 1862. Wounded in action at Fredericksburg, VA, Dec. 13, 1862. Discharged for disability at Portsmouth Grove, RI, Mar. 24, 1863. Died Dec. 7, 1910. Interred at Union Cemetery, North Smithfield, RI.

Brown, John, 2nd. Residence, Providence. 38. M. Laborer. Enlisted Aug. 7, 1862. Mustered in Sept. 6, 1862. Mustered out June 9, 1865.

Brown, John T. Residence, Johnston. 25. S. Express man. Enlisted Aug. 6, 1862. Mustered in Sept. 6, 1862. Discharged for disability at Pleasant Valley, MD, Oct. 25, 1862. Died Dec. 6, 1885. Interred at North Burial Ground, Providence, RI.

Brownell, William. Residence, Johnston. 35. M. Watchman. Enlisted Aug. 5, 1862. Mustered in Sept. 6, 1862. Discharged for disability at Washington, DC, Feb. 24, 1863. Died of disease contracted in the service Mar. 20, 1863 at Johnston, RI. Interred at Intervale Cemetery, North Providence, RI.

Budlong, Benjamin. Residence, Warwick. 26. S. Laborer. Enlisted Aug. 15, 1862. Mustered in Sept. 6, 1862. Mortally wounded in action at Fredericksburg, VA, Dec. 13, 1862. Died of wounds at Washington, DC, Jan. 12, 1863. Interred at Greenwood Cemetery, Coventry, RI.

Burgess, Benjamin W. Residence, Glocester. 33. M. Farmer. Enlisted Aug. 7, 1862. Mustered in Sept. 6, 1862. Killed in action at Fredericksburg, VA, Dec. 13, 1862.

Cady, Nell D. Residence, Glocester. 43. M. Farmer. Enlisted Aug. 6, 1862. Mustered in Sept. 6, 1862. Discharged for disability Nov. 26, 1862. Died Oct. 21, 1897. Interred at Acotes Hill Cemetery, Glocester, RI.

Carr, Alfred D. Residence, North Kingstown. 35. S. Carpenter. Enlisted Aug. 6, 1862, Mustered in Sept. 6, 1862. Wounded in action, shot in neck, at Bethesda Church, VA, June 3, 1864. Absent sick at City Point, VA, Jan. 1865, and so borne until April. Mustered out June 9, 1865.

Carpenter, Richard. Transferred from Co. B, Feb. 1, 1863. Wounded in action at Petersburg, VA, April 2, 1865. Sent to hospital and borne as absent sick at the time his company was mustered out. Mustered out June 30, 1865. Died Jan. 25, 1921. Interred at St. Ann's Cemetery, Cranston, RI.

Case, William S. Transferred from Co. B, Feb. 1, 1865. Deserted May 17, 1865.

Chace, Burrows. Residence, Cranston. 32. M. Mason. Enlisted Aug. 15, 1862. Mustered in Sept. 6, 1862. Discharged for disability at Newport News, VA, Mar. 19, 1863. Died Aug. 20, 1900. Interred at Chace-Weeden Lot, North Kingstown Cemetery 16, North Kingstown, RI.

Clarke, Moses. Residence, Richmond. 32. M. Laborer. Enlisted Aug. 15, 1862. Mustered in Sept. 6, 1862. Mustered out June 9, 1865. Interred at Wood River Cemetery, Richmond, RI.

Collins, Edward F. Transferred from Co. B, Feb. 1, 1865. Mustered out June 9, 1865.

Colvin, Nathan D. Residence, Coventry. 36. S. Farmer. Enlisted Aug. 15, 1862. Mustered in Sept. 6, 1862. Died of typhoid at David's Island, NY Harbor, Sept. 26, 1864. Interred at Royal Colvin Lot, Coventry Cemetery 117, Coventry, RI.

Converse, Marvin J. Residence, Glocester. 32. S. Shoemaker. Enlisted Aug. 7, 1862. Mustered in Sept. 6, 1862. Wounded in action at the Fredericksburg, VA, Dec. 13, 1862. Discharged for disability at Camp Dennison, OH, Oct. 12, 1863, for disability. Died 1918. Interred at Acotes Hill Cemetery, Glocester, RI.

Cook, Lloyd M. Residence, North Providence. 18. S. Painter. Enlisted Aug. 5, 1862. Mustered in Sept. 6, 1862. Mustered out June 9, 1865.

Cornell, Lewis E. Transferred from Co. B, Feb. 1, 1865. Mustered out June 9, 1865. Died April 12, 1901. Interred at Pine Grove Cemetery, Coventry, RI.

Courtney, William. Transferred from Co. B, Feb. 1, 1865. Mustered out June 9, 1865.

Cutler, Samuel B. Residence, Glocester. 28. S. Farmer. Enlisted Aug. 5, 1862. Mustered in Sept. 6, 1862. Absent sick Nov. 1862. Discharged for disability at General Hospital Dec. 20, 1862.

Daniels, Herbert. Transferred from Co. B, Feb. 1, 1865. Mustered out June 9, 1865. Died Nov. 26, 1899. Interred at North Burial Ground, Providence, RI.

Dean, Isaac N. Transferred from Co. B, Co. B, Feb. 1, 1865. Mustered out June 9, 1865. Died Sept. 20, 1895. Interred at Togus National Cemetery. Grave 1218.

Doherty, Bernard. Residence, North Providence. 18. S. Laborer. Enlisted and Mustered Feb. 14, 1865. Transferred to Co. B, June 9, 1865.

Dorrance, John. Residence, Foster. 22. S. Stone cutter. Enlisted Aug. 5, 1862. Mustered in Sept. 6, 1862. Died of typhoid in Second Division, Ninth Army Corps Field Hospital, Windmill Point, at Aquia Creek, VA, Jan. 26, 1863. Interred at Benjamin Cahoone Lot, Coventry Cemetery 101, Coventry, RI.

Dugan, Hugh. Transferred from Co. B, Feb. 1, 1865. Mustered out June 9, 1865. Interred at Togus National Cemetery. Grave 1247.

Durfee, Gilbert. Residence, Glocester. 25. S. Farmer. Enlisted Aug. 7, 1862. Mustered in Sept. 6, 1862. Killed in action at Poplar Spring Church, VA, Sept. 30, 1864. Interred at Smithville Cemetery, Scituate, RI.

Eddy, John H. Residence, Glocester. 22. M. Farmer. Enlisted Aug. 7, 1862. Mustered in Sept. 6, 1862. Wounded in action, shot in right hand, at Fredericksburg, VA, Dec. 13, 1862. Wounded in action, shot in leg and amputated, July 1, 1864, at Petersburg, VA. Sent to hospital. Discharged for disability at Portsmouth Grove, RI, Feb. 1, 1865. Died Nov. 27, 1876. Interred at Acotes Hill Cemetery, Glocester, RI

Eldridge, James E. Residence, Warwick. 19. S. Farmer. Enlisted Aug. 15, 1862. Mustered in Sept. 6, 1862. Died of typhoid at Washington, DC, July 16, 1864.

Farnum, Edwin A. Transferred from Co. B, Feb. 1, 1865. Mustered out June 9, 1865. Died Mar. 9, 1913. Interred at Acotes Hill Cemetery, Glocester, RI.

Flemming, Thomas. Transferred from Co. B, Feb. 1, 1865. Mustered out June 9, 1865. Died January 9, 1916. Interred at Togus National Cemetery. Grave 3387.

Foley, Dennis. Transferred from Co. B, Feb. 1, 1865. Mustered out June 9, 1865. Died May 2, 1887. Interred at Togus National Cemetery. Grave 596.

Gardiner, Joseph W. Residence, East Greenwich. 25. S. Blacksmith. Enlisted Aug. 5, 1862. Mustered in Sept. 6, 1862. Absent sick at Washington, DC, Jan. 1865, and so borne until April 1865. Mustered out at Washington, DC, May 25, 1865. Died July 16, 1900. Interred at Brayton Cemetery, Warwick, RI.

Goldwaite, George E. Residence, Portsmouth. 21. S. Machinist. Enlisted Aug. 8, 1862. Mustered in Sept. 6, 1862. Transferred to the Veteran Reserve Corps Jan. 16, 1864.

Greene, Daniel. Residence, North Kingstown. 40. M. Laborer. Enlisted Aug. 7, 1862. Mustered in Sept. 6, 1862. Wounded in action, shot in spine, at Fredericksburg, VA, Dec. 13, 1862. Discharged for disability Jan. 17, 1863. Died Jan. 26, 1902. Interred at Elm Grove Cemetery, North Kingstown, RI.

Greene, William. Residence, Providence. 42. M. Mason. Enlisted Aug. 15, 1862. Mustered in Sept. 6, 1862. Transferred to Co. A as musician Oct. 3, 1862.

Harrah, Matthew. Residence, Newport. 24. S. Dresser. Enlisted Aug. 12, 1862. Mustered in Sept. 6, 1862. Wounded in action, in head, at Fredericksburg, VA, Dec. 13, 1862. Appointed Division postmaster Jan. 1865, and so borne until June 1865. Mustered out June 9, 1865.

Harrington, Albert. Residence, Warwick. 40. M. Laborer. Enlisted Aug. 15, 1862. Mustered in Sept. 6, 1862. Wounded in action, shot in leg, at Petersburg, VA, Aug. 9, 1864. Mustered out June 9, 1865. Died Dec. 10, 1899. Interred at Brayton Cemetery, Warwick, RI.

Hatfield, Richard. Residence, Glocester. 47. M. Printer. Enlisted Aug. 8, 1862. Mustered in Sept. 6, 1862. Died of typhoid at Alexandria, VA, Nov. 19, 1862. Interred at Alexandria National Cemetery. Grave 450. Cenotaph at North Burial Ground, Providence, RI.

Holland, Charles W. Residence, South Kingstown. 28. S. Spinner. Enlisted Aug. 7, 1862. Mustered in Sept. 6, 1862. Wounded in action at the Wilderness, May 5, 1864. In division hospital Jan. 1865, and so borne until June 1865. Mustered out June 9, 1865. Died Feb. 23, 1892. Interred at Riverside Cemetery, South Kingstown, RI.

Keech, Benjamin A. Residence, Smithfield. 34. S. Painter. Enlisted Aug. 7, 1862. Mustered in Sept. 6, 1862. Wounded in action, shot in hip, at Petersburg, VA, June 29, 1864. Discharged for disability at Portsmouth Grove, RI, June 1, 1865.

Killian, John H. Residence, Portsmouth. 29. S. Laborer. Enlisted Aug. 8, 1862. Mustered in Sept. 6, 1862. Wounded in action, shot in wrist, at Bethesda Church, VA, June 3, 1864. Discharged for disability at Willett's Point, NY Harbor, Dec. 8, 1864. Interred at St. Mary's Cemetery, Newport, RI.

Knight, Alfred Sheldon. Residence, Scituate. 29. S. Farmer. Enlisted Aug. 16, 1862. Mustered in Sept. 6, 1862. Died of pneumonia in regimental hospital at Falmouth, VA, Jan. 31, 1863. Interred at W.W. Knight Lot, Scituate Cemetery 76, Scituate, RI.

Lamby, Peter. Transferred from Co. B, Feb. 1, 1865. Mustered out June 9, 1865.

Langland, Isaac. Residence, Bristol. 18. S. Laborer. Enlisted and Mustered Feb. 17, 1865. Transferred to Co. B, June 9, 1865.

Laugherty, John. Transferred from Co. B, Feb. 1, 1865. Mustered out June 9, 1865.

Lawton, Alfred. Residence, Exeter. 31. M. Spinner. Enlisted Aug. 1, 1862. Discharged for disability at Camp Dennison, OH, Oct. 28, 1863. Died 1908. Interred at Knotty Oak Cemetery, Coventry, RI.

Lawton, Joseph S. C. Residence, Glocester. 43. M. Mason. Enlisted Aug. 6, 1862. Mustered in Sept. 6, 1862. Wounded in action, shot in side, at Bethesda Church, VA, June 3, 1864. Wounded in action April 2, 1865 at Petersburg, VA. Mustered out June 9, 1865. Died Jan. 16, 1876. Interred at Acotes Hill Cemetery, Glocester, RI.

McCafferey, Patrick. Transferred from Co. B, Feb. 1, 1865. Discharged for disability at Fort Sedgwick, VA, Mar. 4, 1865.

McCready, Daniel. Transferred from Co. B, Feb. 1, 1865. Wounded in action, shot in side, at Petersburg, VA, April 2, 1865. Mustered out June 9, 1865.

McDermott, John. Residence, Glocester. 49. M. Farmer. Enlisted Aug. 7, 1862. Mustered in Sept. 6, 1862. Borne as sick at Washington Dec. 20, 1862, until Feb. 1863. Discharged for disability at Baltimore, MD, Mar. 14, 1863. Died May 1, 1865. Interred at Pascoag Cemetery, Burrillville, RI.

McDonald, John J. Residence, Glocester. 40. M. Machinist. Enlisted Aug. 11, 1862. Mustered in Sept. 6, 1862. Wounded in action at Cold Harbor, VA, June 4, 1864. Mustered out June 9, 1865. Died of disease contracted in the service at Providence, RI, Aug. 29, 1866.

McLaughlin, Neil. Transferred from Co. B, Feb. 1, 1865. Mustered out June 9, 1865. Died June 18, 1905. Interred at Boston, MA.

Marant, Elisha A. Transferred from Co. B, Feb. 1, 1865. Mustered out June 9, 1865.

Matteson, Calvin R. Residence, West Greenwich. 21. S. Farmer. Enlisted Aug. 7, 1862. Mustered in Sept. 6, 1862. Wounded in action at Fredericksburg, VA, Dec. 13, 1862. Discharged for disability at Portsmouth Grove, RI, April 2, 1863. Lost at sea Dec. 22, 1864 onboard the *North America* returning home from war, having reenlisted in Troop G, 3rd Rhode Island Cavalry. Cenotaph at Large Maple Root Cemetery, Coventry, RI.

Matthewson, Cornelius. Transferred from Co. B, Feb. 1, 1865. Mustered out June 9, 1865.

Miller, Francis B. Residence, Glocester. 26. M. Dresser. Enlisted Aug. 11, 1862. Mustered in Sept. 6, 1862. Absent sick at Washington, DC, Jan. 1865. Mustered out June 9, 1865. Died Feb. 6, 1880. Interred at Woodlawn Cemetery, Johnston, RI.

Moran, Patrick. Transferred from Co. B, Feb. 1, 1865. Mustered out July 31, 1865. Died Aug. 15, 1913. Interred at St. James Cemetery, Danielson, CT.

Mulvey, Michael. Transferred from Co. A, Oct. 1862. Mustered out June 9, 1865.

Neville, Edwin M. Residence, Johnston. 22. S. Farmer. Enlisted Aug. 5, 1862. Mustered in Sept. 6, 1862. Discharged for disability at Falmouth, VA, Dec. 9, 1862.

Norton, Sylvester E. Transferred from Co. B, Feb. 1, 1865. Transferred to Veteran Reserve Corps in Feb. 1865.

Nye, Byron C. Transferred from Co. B, Feb. 1, 1865. Mustered out June 9, 1865. Died March 5, 1922. Interred at Togus National Cemetery. Grave 3872.

Oliver, Arthur. Residence, Providence. 35. S. Spinner. Enlisted Aug. 18, 1862. Mustered in Sept. 6, 1862. Mustered out June 9, 1865.

Orcutt, Albert G. Residence, Portsmouth. 25. S. Moulder. Enlisted Aug. 8, 1862. Mustered in Sept. 6, 1862. Transferred to Veteran

Reserve Corps, Jan. 15, 1864. Died Feb. 28, 1902. Interred at Pocasset Cemetery, Cranston, RI.

Quinlan, Thomas. Transferred from Co. B, Feb. 1, 1865. Mustered out June 9, 1865.

Owens, Thomas T. Residence, Warwick. 20. S. Laborer. Enlisted and Mustered Jan. 24, 1865. Detached as orderly at brigade headquarters June 1865. Transferred to Co. B, June 9, 1865.

Page, Harlan A. Residence, Glocester. 20. S. Farmer. Enlisted Aug. 7, 1862. Mustered in Sept. 6, 1862. On duty at brigade headquarters from Jan. 1865 to June, 1865. Mustered out June 9, 1865. Died Oct. 22, 1926. Interred at Pocasset Cemetery, Cranston, RI.

Paine, Daniel W. Residence, Glocester. 28. M. Farmer. Enlisted Aug. 7, 1862. Mustered in Sept. 6, 1862. Mustered out June 9, 1865. Died Dec. 21, 1909. Interred at Acotes Hill Cemetery, Glocester, RI.

Platt, Thomas W. Residence, Pawtucket. 25. S. Weaver. Enlisted Aug. 7, 1862. Mustered in Sept. 6, 1862. Discharged for disability Mar. 22, 1863. Died Mar. 23, 1902. Interred at Greenwood Cemetery, Coventry, RI.

Potter, Francis W. Residence, Cranston. 34. M. Blacksmith. Enlisted Aug. 7, 1862. Mustered in Sept. 6, 1862. Promoted to corporal.

Radigan, James I. Residence, Warwick. 21. S. Operative. Enlisted Aug. 6, 1862. Mustered in Sept. 6, 1862. Wounded in action at Fredericksburg, VA, Dec. 13, 1862. Deserted at Lexington, KY, April 1, 1863.

Ratcliffe, Richard. Residence, Providence. 30. M. Weaver. Enlisted Aug. 18, 1862. Mustered in Sept. 6, 1862. Killed in action at Fredericksburg, VA, Dec. 13, 1862.

Riley, John. Transferred from Co. B, Feb. 1, 1865. Mustered out June 9, 1865.

Reardon, Edward. Residence, Providence. 36. M. Bootmaker. Enlisted Aug. 20, 1862. Mustered in Sept. 6, 1862. Wounded in action at Petersburg, VA, April 2, 1865. Discharged for disability at Carver Hospital, Washington, DC, June 12, 1865.

Robbins, Nathan N. Residence, Johnston. 18. S. Spinner. Enlisted Aug. 5, 1862. Mustered in Sept. 6, 1862. Died of Yazoo Fever at Big Black River, MS, July 22, 1863. Interred at Vicksburg National Cemetery. Section Q, Grave 28.

Rourke, Patrick. Residence, Burrillville. 28. S. Farmer. Enlisted July 30, 1862. Mustered in Sept. 6, 1862. Mortally wounded in action, shot in scalp, Dec. 13, 1862 at Fredericksburg, VA. Sent to Baltimore, MD. Deserted from hospital at Baltimore, MD, April 30, 1863 and returned to Burrillville, RI. Died of wounds at Burrillville, RI July 8, 1863. Interred at Pascoag Cemetery, Burrillville, RI.

Saunders, George A. Residence, Glocester. 21. S. Farmer. Enlisted Aug. 8, 1862. Mustered in Sept. 6, 1862. Discharged for disability at Newport News, VA, Mar. 19, 1863. Interred at Acotes Hill Cemetery, Glocester, RI.

Sayles, Lemuel C. Residence, Glocester. 18. S. Laborer. Enlisted Aug. 8, 1862. Mustered in Sept. 6, 1862. Transferred to Veteran Reserve Corps, Sept. 1, 1863. Died Sept. 27, 1898. Interred at Pascoag Cemetery, Burrillville, RI.

Schouler, James. Residence, Johnston. 41. M. Weaver. Enlisted Aug. 10, 1862. Mustered in Sept. 6, 1862. Discharged for disability at Fort Columbus, NY, July 27, 1863.

Seaver, William H. Residence, Providence. 23. S. Blacksmith. Enlisted July 14, 1862. Mustered in Sept. 6, 1862. Discharged for disability at Newport News, VA, Mar. 2, 1863. Died June 17, 1865. Interred at Grace Church Cemetery, Providence, RI.

Sheridan, John. Transferred from Co. B, Feb. 1, 1865. Mustered out June 9, 1865.

Sherman, David. Residence, Pawtucket. 18. S. Laborer. Enlisted and Mustered Mar. 19, 1865. Transferred to Co. B, June 9, 1865.

Shippee, Justin. Residence, Foster. 23. S. Farmer. Enlisted Aug. 8, 1862. Mustered in Sept. 6, 1862. Transferred to Veteran Reserve Corps, May 6, 1864.

Smith, Edwin R. Residence, Providence. 18. S. Laborer. Enlisted and Mustered in Mar. 2, 1865. Mustered out July 13, 1865. Interred at Greenville Cemetery, Smithfield, RI.

Smith, William H. Residence, Smithfield. 30. M. Carpenter. Enlisted Aug. 7, 1862. Mustered in Sept. 6, 1862. Jan. 1863, at division headquarters. Mar. 1865, Ambulance Corps, and so borne until June 1865. Mustered out June 9, 1865.

Spencer, Edwin C. Residence, East Greenwich. 18. S. Farmer. Enlisted Aug. 7, 1862. Mustered in Sept. 6, 1862. Mustered out June 9, 1865. Died Aug. 2, 1923. Interred at First Cemetery, East Greenwich, RI.

Steadman, William H. Residence, Providence. 24. S. Laborer. Enlisted and Mustered Mar. 1, 1865. Transferred to Co. B, June 9, 1865.

Steere, Henry. Residence, Glocester. 18. S. Farmer. Enlisted Aug. 6, 1862. Mustered in Sept. 6, 1862. Transferred to Veteran Reserve Corps, Jan. 15, 1864. Died Sept. 9, 1897. Interred at Dexter Lot, Scituate Cemetery 8, Scituate, RI.

Steere, Horatio. Residence, Smithfield. 42. M. Trader. Enlisted Aug. 5, 1862. Mustered in Sept. 6, 1862. Discharged for disability at Pleasant Valley, MD, Oct. 25, 1862. Died 1890. Interred at Moshassuck Cemetery, Central Falls, RI.

Sweetland, Job R. Residence, Pawtucket. 21. S. Tailor. Enlisted Aug. 7, 1862. Mustered in Sept. 6, 1862. Mortally wounded in action at Fredericksburg, VA, Dec. 13, 1862. Died of wounds at Washington, DC, Feb. 27, 1863. Interred at Mineral Spring Cemetery, Pawtucket, RI.

Taft, Isaac J. Residence, Burrillville. 33. M. Carpenter. Enlisted Aug. 8, 1862. Mustered in Sept. 6, 1862. Mustered out June 9, 1865. Died June 6, 1884. Interred at Acotes Hill Cemetery, Glocester, RI.

Thornley, Richard. Residence, Newport. 19. S. Laborer. Enlisted and Mustered Feb. 21, 1865. Transferred to Co. B, June 9, 1865.

Tuckerman, James F. Residence, Warwick. 28. M. Laborer. Enlisted Aug. 7, 1862. Mustered in Sept. 6, 1862. Detached to Battery D, 1st RI Light Artillery, Jan. 15, 1863, and so borne until Dec. 10, 1864, when he was transferred to the 7th RI Vols. Mustered out June 9, 1865. Died May 14, 1878. Interred at Knotty Oak Cemetery, Coventry, RI.

Turner, James. Residence, Newport. 30. M. Weaver. Enlisted Aug. 13, 1862. Mustered in Sept. 6, 1862. Discharged for disability at Pleasant Valley, MD, Oct. 25, 1862. Died Mar. 31, 1911. Interred at Moshassuck Cemetery, Central Falls, RI.

Turner, Thomas. Residence, Newport. 28. M. Weaver. Enlisted Aug. 13, 1862. Mustered in Sept. 6, 1862. Wounded in action, shot in hand, at Spotsylvania Court House, VA, May 12, 1864. Discharged for disability at Alexandria, VA, May 8, 1865.

Vallet, Jedediah S. Residence, Burrillville. 21. S. Laborer. Enlisted Aug. 8, 1862. Mustered in Sept. 6, 1862. Transferred to Veteran Reserve Corps, Jan. 15, 1864. Died Dec. 28, 1904. Interred at Acotes Hill Cemetery, Glocester, RI.

Weldon, George W. Residence, Providence. 21. S. Laborer. Enlisted and Mustered Mar. 21, 1865. Transferred to Co. B, June 9, 1865.

White, Elijah F. Residence, Providence. 27. M. Express man. Enlisted Aug. 28, 1862. Mustered in Sept. 6, 1862. Mustered out

June 9, 1865. Died Oct. 3, 1898. Interred at North Burial Ground, Providence, RI.

Williams, Edwin P. Residence, North Providence. Deserter from 12th RI Vols. Sent to serve out term with 7th Rhode Island and taken up on roll Dec. 13, 1863. Mustered out Sept. 20, 1864. Died Dec. 30, 1906. Interred at Greenville Cemetery, Smithfield, RI.

Wood, Daniel. Residence, Portsmouth. 30. M. Mechanic. Enlisted Aug. 8, 1862. Mustered in Sept. 6, 1862. Absent sick at Portsmouth Grove, RI, Jan. 1865, and so borne until June 1865. Mustered out June 20, 1865. Died 1911. Interred at Greenwood Cemetery, Coventry, RI.

Young, Emor. Residence, Glocester. 39. M. Lumberman. Enlisted Aug. 7, 1862. Mustered in Sept. 6, 1862. Wounded in action at Petersburg, VA, April 2, 1865. Mustered out June 9, 1865. Died of disease contracted in the service at Glocester, RI, Nov. 20, 1868. Interred at Acotes Hill Cemetery, Glocester, RI.

COMPANY D

Captain

Channell, Alfred M. Promoted from first lieutenant Co. G, Oct. 24, 1862, and assumed command Dec. 1, 1862. Dismissed from the service for drunkenness on duty and conduct unbecoming an officer Aug. 1, 1864. Died Aug. 20, 1884. Interred at Hope Cemetery, Alexis Junction, IL.

First Lieutenants

Joyce, William H. Residence, Providence. 26. M. Book keeper. Commissioned Sept. 4, 1862. Mustered in Sept. 6, 1862. Promoted to captain Co. F, Jan. 7, 1862.

Merrill, James F. Promoted from second lieutenant Co. C, Oct. 31, 1863. Transferred to Co. I, Feb. 1, 1865.

Smith, Albert L. Promoted from second lieutenant Co. I, April 3, 1863, and assigned to Co. D, May 19, 1863. Died of Yazoo Fever at Nicholasville, KY, Aug. 31, 1863. Interred at Camp Nelson National Cemetery. Section D, Grave 1260.

Second Lieutenants

Dingley, Fuller. Promoted from sergeant Co. I, May 20, 1863. Captured at Jackson, MS, July 13, 1863. Borne as a prisoner of war until Nov. 1864. Exchanged. Mustered out Mar. 23, 1865. Died Nov. 18, 1897. Interred at Oak Grove Cemetery, Gardiner, ME.

Hathaway, Cyrus B. Residence, Pawtucket. 30. S. Jeweler. Commissioned May 26, 1862. Mustered in Sept. 6, 1862. Commissioned first lieutenant Jan. 7, 1863, but never mustered as such. Resigned Jan. 13, 1863. Died 1879. Interred at Swan Point Cemetery, Providence, RI.

First Sergeants

Congdon, George W. Promoted from corporal Jan. 7, 1863. Killed in action at Bethesda Church, VA, June 3, 1864. Interred at Cold Harbor National Cemetery. Grave 806.

Johnson, William H. Promoted from private June 29, 1864. Commissioned second lieutenant July 25,1864, but never mustered as such. Promoted to first lieutenant Co. E, Oct. 21, 1864.

Sullivan, John. Residence, Smithfield. Enlisted Aug. 1, 1862. 29. S. Soldier. Mustered in Sept. 6, 1862. On detached duty as acting sergeant major, Dec. 1862. Promoted to second lieutenant Co. K, Jan. 7, 1863.

Sergeants

Bolles, Albert A. Promoted from corporal. Promoted to second lieutenant Co. F, Mar. 1, 1863.

Harrington, William. Promoted from private. Died of dysentery at hospital, Nicholasville, KY, Aug. 31, 1863. Interred at Camp Nelson National Cemetery. Section L, Grave 1261.

Lowell, John Z. Residence, Tiverton. 26. M. Grocer. Enlisted Aug. 13, 1862. Mustered in Sept. 6, 1862. Discharged for disability at Newport News, VA, Mar. 3, 1863. Died July 9, 1865. Interred at Mt. Hope Cemetery, Boston, MA.

Merrill, James F. Residence, Providence. 24. S. Clerk. Enlisted July 30, 1862. Mustered in Sept. 6, 1862. Promoted to second lieutenant Co. C, Mar. 1, 1863.

Sanderson, Edward. Residence, Providence. 29. M. Machinist. Enlisted Feb. 7, 1862. Borne as absent sick from Oct. 3, 1862, until Nov. 12, 1862, when he deserted at hospital at Baltimore, MD.

Sprague, John H. D. Promoted from corporal. Wounded in action, at Petersburg, VA, June 29, July 1, and July 7, 1864. In

Ambulance Corps, Jan. 1865, and so borne until June 1865. Transferred to Co. I, Feb. 1, 1865.

Young, Henry. Residence, Providence. 35. M. Grocer. Enlisted Aug. 8, 1862. Mustered in Sept. 6, 1862. Promoted to second lieutenant Co. H, Mar. 1, 1863.

Corporals

Bolles, Albert A. Residence, Pawtucket. 29. M. Marble worker. Enlisted Aug. 14, 1862. Mustered in Sept. 6, 1862. Promoted to sergeant.

Case, William H. Promoted from private. Discharged for disability at Newport News, VA, Mar. 3, 1863.

Congdon, George W. Residence, Pawtucket. 21. M. Jeweler. Enlisted Aug. 12, 1862. Mustered in Sept. 6, 1862. Promoted to first sergeant Jan. 7, 1863.

Corey, Alvin L. Residence, Coventry. 31. M. Farmer. Enlisted Aug. 6, 1862. Mustered in Sept. 6, 1862. Discharged for disability Jan. 1, 1863 at Falmouth, VA. Died 1911. Interred at New Westfield Cemetery, Danielson, CT.

Darling, Esek R. Residence, Burrillville. 19. M. Farmer. Enlisted July 29, 1862. Mustered in Sept. 6, 1862. Wounded in action, shot in finger, at Fredericksburg, VA, Dec. 13, 1862. Transferred to Co. I, Feb. 1, 1865.

McAdams, James. Promoted from private Sept. 11, 1862. Deserted while marching through Baltimore, MD, Sept. 12, 1862.

Potter, Henry C. Residence, North Providence. 45. M. Harness maker. Enlisted Aug. 8, 1862. Mustered in Sept. 6, 1862. Discharged for disability at Washington, DC, Nov. 24, 1862. Died February 28, 1906. Interred at Swan Point Cemetery, Providence, RI.

Sherman, Daniel B. Promoted from private. Killed in action at Spotsylvania Court House, VA, May 18, 1864. Served in Color Guard. Cenotaph at First Cemetery, East Greenwich, RI.

Shumway, Amos D. Residence, Burrillville. 19. S. Shoemaker. Enlisted July 30, 1862. Mustered in Sept. 6, 1862. Transferred to Co. I Feb. 1, 1865.

Sprague, John H. D. Residence, Burrillville. 21. S. Farmer. Enlisted July 29, 1862. Mustered in Sept. 6, 1862. Promoted to sergeant.

Thompson, Elisha E. Residence, Burrillville. 20. S. Hostler. Enlisted July 31, 1862. Mustered in Sept. 6, 1862. Wounded in action, shot in thumb, at Fredericksburg, VA, Dec. 13, 1862. Discharged for disability at Portsmouth Grove, RI, Mar. 19, 1863. Died June 15, 1905. Interred at South Street Cemetery, Douglas, MA.

Musicians

Hopkins, Charles W. Residence, West Greenwich. 23. M. Teacher. Enlisted Aug. 14, 1862. Mustered in Sept. 6, 1862. Borne as absent sick at Washington from Nov. 17, 1862, until Feb. 1, 1863. Clerk in brigade commissary department Jan. 1865, and so borne until May 1865. Transferred to Co. A, Feb. 1, 1865.

Hopkins, William P. Residence, West Greenwich. 17. S. Carpenter. Enlisted Aug. 14, 1862. Mustered in Sept. 6, 1862. Transferred to Co. I, Feb. 1, 1865.

Wagoner

Branigan, John B. Residence, Burrillville. 19. S. Farmer. Enlisted July 30, 1862. Mustered in Sept. 6, 1862. On detached service at Fort Wood, New York Harbor, Jan. 1865, and so borne until June 1865. Transferred to Co. I, Feb. 1, 1865.

Privates

Beckford, George C. Residence, Providence. 29. M. Moulder. Enlisted Aug. 7, 1862. Mustered in Sept. 6, 1862. Transferred to Co. I, Feb. 1, 1865.

Beaumont, Ralph. Residence, Burrillville. 21. S. Shoemaker. Enlisted July 29, 1862. Mustered in Sept. 6, 1862. Detached to Ambulance Corps Jan. 1863. Transferred to Co. I, Feb. 1, 1865.

Bradbury, John. Residence, Coventry. 24. S. Mule spinner. Enlisted Aug. 11, 1862. Mustered in Sept. 6, 1862. Wounded in action, shot in foot, at Fredericksburg, VA, Dec. 13, 1862. Sent to Washington and borne as absent sick from that time until Feb. 1863. Transferred to Co. I, Feb. 1, 1865.

Brennan, John. Residence, Smithfield. 38. S. Painter. Enlisted Aug. 8, 1862. Mustered in Sept. 6, 1862. Wounded in action, shot in hand, at Fredericksburg, VA, Dec. 13, 1862. Discharged for disability at Portsmouth Grove, RI, Aug. 6, 1864.

Brown, Marcus M. Residence, Burrillville. 19. S. Farmer. Enlisted Aug. 8, 1862. Mustered in Sept. 6, 1862. Transferred to Co. I, Feb. 1, 1865.

Bryden, Wilson C. Residence, Burrillville. 18. S. Shoemaker. Enlisted Aug. 6, 1862. Mustered in Sept. 6, 1862. Transferred to Co. K, Oct. 14, 1862.

Bullard, Dwight J. Residence, Greenfield, MA. 39. S. Laborer. Enlisted Aug. 9, 1862. Mustered in Sept. 6, 1862. Transferred to Veteran Reserve Corps, May 1, 1864.

Bullock, Allen E. Residence, North Kingstown. 16. S. Sailor. Enlisted Aug. 9, 1862. Mustered in Sept. 6, 1862. Wounded in action, shot in side, at Petersburg, VA, June 26, 1864. Discharged for disability, at Fort Sedgwick, VA, Jan. 6, 1865. Died Nov. 28, 1911. Interred at Elm Grove Cemetery, North Kingstown, RI.

Case, William H. Residence, Providence. 37. M. Laborer. Enlisted Aug. 13, 1862. Mustered in Sept. 6, 1862. Promoted to corporal.

Carragan, Martin W. Residence, Portsmouth. 21. S. Weaver. Enlisted Aug. 9, 1862. Wounded in action at Fredericksburg, VA, Dec. 13, 1862. Wounded in action, shot in ankle, at Bethesda Church, VA, June 3, 1864. Transferred to Co. I, Feb. 1, 1865.

Callahan, Timothy A. Residence, Boston, MA. 40. M. Carpenter. Enlisted Aug. 7, 1862. Mustered in Sept. 6, 1862. Deserted while marching through Baltimore, MD, Sept. 12, 1862.

Brown, Charles H. Residence, Warwick. 22. S. Farmer. Enlisted Aug. 7, 1862. Mustered in Sept.6, 1862. Discharged for disability at Newport News, VA, Mar. 3, 1863. Died 1925. Interred at Brayton Cemetery, Warwick, RI.

Chace, Joseph. Residence, North Kingstown. 44. M. Machinist. Enlisted Aug. 8, 1862. Mustered in Sept. 6, 1862. Borne as absent sick in hospital from Aug. 30, 1864, until Dec. 29, 1864, when he was transferred to the Veteran Reserve Corps. Died April 4, 1887. Interred at Togus National Cemetery. Grave 590.

Daggett, Seril N. Residence, Glocester. 42. M. Mason. Enlisted Aug. 4, 1862. Mustered in Sept. 6, 1862. Wounded in action, shot in left knee, at Fredericksburg, VA, Dec. 13, 1862, and sent to hospital. Discharged for disability Aug. 5, 1864. Died June 1, 1878. Interred at Acotes Hill Cemetery, Glocester, RI.

Danforth, George A. Residence, Providence. 29. S. Clerk. Enlisted Aug. 8, 1862. Mustered in Sept. 6, 1862. Transferred to Co. I, Feb. 1, 1865.

Dawley, Varnum H. Residence, Exeter. 19. M. Machinist. Enlisted Aug. 9, 1862. Mustered in Sept. 6, 1862. Transferred to Co. I, Feb. 1, 1865.

Denicoe, Frank Jr. Residence, Providence. 28. M. Farmer. Enlisted Aug. 8, 1862. Mustered in Sept. 6, 1862. Wounded in action, finger shot away, at Fredericksburg, VA, Dec. 13, 1862. Wounded in action, shot in hand, at Bethesda Church, VA, June 3, 1864. Transferred to Co. I, Feb. 1, 1865.

Denicoe, Joseph. Residence, North Kingstown. 17. S. Farmer. Enlisted Aug. 9, 1862. Mustered in Sept. 6, 1862. Wounded in action at Fredericksburg, VA, Dec. 13, 1862. Transferred to Co. I, Feb. 1, 1865.

Donnelly, Patrick. Residence, Providence. 44. S. Laborer. Enlisted Aug. 2, 1862. Mustered in Sept. 6, 1862. Died of dysentery in hospital at Lexington, KY, June 30, 1863. Interred at St. Francis Cemetery, Pawtucket, RI.

Durfee, Albert G. Residence, Smithfield. 34. S. Carpenter. Enlisted Aug. 11, 1862. Mustered in Sept. 6, 1862. Discharged for disability at Falmouth, VA, Feb. 24, 1863. Died Jan. 2, 1889. Interred at Swan Point Cemetery, Providence, RI.

Fagan, Patrick. Residence, Burrillville. 31. M. Laborer. Enlisted July 31, 1862. Mustered in Sept. 6, 1862. Transferred to Co. I, Feb. 1, 1865.

Frowley, John. Residence, Providence. 25. S. Sailor. Enlisted Aug. 9, 1862. Mustered in Sept. 6, 1862. Honorably discharged Oct. 24, 1862, and enlisted Oct. 25, 1862, in Battery E, 4th United States Artillery.

Gorton, Burrill B. Residence, West Greenwich. 18. S. Farmer. Enlisted Aug. 12, 1862. Mustered in Sept. 6, 1862. Discharged for disability at Falmouth, VA, Dec. 10, 1862. Died 1889. Interred at Spears Cemetery, Foster, RI.

Hackett, James. Residence, Providence. 18. S. Sailor. Enlisted Aug. 8, 1862. Mustered in Sept. 6, 1862. Wounded in action at Fredericksburg, VA, Dec. 13, 1862. Discharged for disability at Portsmouth Grove, RI, Mar. 20, 1863.

Harrington, Russell. Residence, Burrillville. 40. S. Farmer. Enlisted July 30, 1862. Mustered in Sept. 6, 1862. Discharged for disability Nov. 30, 1862.

Harrington, William. Residence, Scituate. 21. S. Laborer. Enlisted Aug. 6, 1862. Mustered in Sept. 6, 1862. Promoted to sergeant.

Harris, Jeremiah. Residence, Burrillville. 38. M. Sailor. Enlisted Aug. 2, 1862. Mustered in Sept. 6, 1862. Deserted at Washington, DC, Sept. 28, 1862.

Hayes, Samuel A. Residence, Little Compton. 36. M. Mule spinner. Enlisted Aug. 11, 1862. Mustered in Sept. 6, 1862. Transferred to Co. I, Feb. 1, 1865.

Holley, William A. Residence, North Kingstown. 16. S. Farmer. Enlisted Aug. 8, 1862. Mustered in Sept. 6, 1862. Wounded in action at Poplar Spring Church, VA, Sept. 30, 1864. Transferred to Veteran Reserve Corps, Jan. 10, 1865. Interred at Elm Grove Cemetery, North Kingstown, RI.

Humes, Charles H. Residence, Burrillville. 18. S. Farmer. Enlisted, Aug. 6, 1862. Mustered in Sept. 6, 1862. Wounded in action, shot in hand, at Cold Harbor, VA, June 6, 1864. Transferred to Co. I, Feb. 1, 1865.

Humes, Emory. Residence, Burrillville. 19. S. Shoemaker. Enlisted Aug. 2, 1862. Mustered in Sept. 6, 1862. Transferred to Co. I, Feb. 1, 1865.

Irons, Charles A. S. Residence, Johnston. 37. M. Spinner. Enlisted July 30, 1862. Mustered in Sept. 6, 1862. Borne as absent sick from Sept. 29, 1864, until April 1865. Transferred to Co. I, Feb. 1, 1865.

Johnson, William H. Transferred from Co. I, Jan. 29, 1863. Promoted to first sergeant June 29, 1863.

Joslin, Benjamin. Residence, Burrillville. 21. S. Hostler. Enlisted July 29, 1862. Mustered in Sept. 6, 1862. Transferred to Co. I, Feb. 1, 1865.

Keogh, James. Residence, Pawtucket. 40. M. Laborer. Enlisted Aug. 12, 1862. Mustered in Sept. 6, 1862. Transferred to Veteran Reserve Corps September 6, 1864.

Kerr, Michael. Residence, Pawtucket. 18. S. Laborer. Enlisted Aug. 9, 1862. Mustered in Sept. 6, 1862. Wounded in action, shot in head, at Fredericksburg, VA, Dec. 13, 1862. Discharged for disability at Portsmouth Grove, RI, Feb. 1, 1863.

Lee, Cornelius. Residence, Boston, MA. 19. S. Book binder. Enlisted Aug. 5, 1862. Mustered in Sept. 6, 1862. Transferred to Co. I, Feb. 1, 1865.

Locklin, Thomas, Jr. Residence, North Providence. 18. S. Operative. Enlisted Feb. 15, 1864. Transferred to Co. I, Feb. 1, 1865.

McAdams, James. Residence, Portsmouth. 23. S. Machinist. Enlisted Aug. 7, 1862. Mustered in Sept. 6, 1862. Promoted to corporal Sept. 11, 1862.

McKenna, Owen. Residence, Pawtucket. 32. M. Laborer. Enlisted Aug. 15, 1862. Mustered in Sept. 6, 1862. Killed in action at Spotsylvania Court House, VA, May 18, 1864. Cenotaph at St. Mary's Cemetery, Pawtucket, RI.

McNaulty, Hugh. Residence, Providence. 44. M. Mason. Enlisted Aug. 11, 1862. Mustered in Sept. 6, 1862. Transferred to Co. I, Feb. 1, 1865.

McQueeny, Barnard. Residence, Boston, MA. 37. M. Grocer. Enlisted Aug. 5, 1862. Mustered in Sept. 6, 1862. Deserted Sept. 10, 1862. Joined regiment from desertion July 10, 1864, and sentenced to three years hard labor at Tortugas, FL, by general court-martial. Died of yellow fever at Fort Jefferson, FL, Aug. 6, 1865.

Minz, John. Residence, Providence. Enlisted Aug. 4, 1862. Mustered in Sept. 6, 1862. Deserted at Baltimore, MD, Mar. 27, 1863.

Murray, Frank. Residence, Woonsocket. 38. M. Tailor. Enlisted Aug. 8, 1862. Mustered in Sept. 6, 1862. Transferred to Veteran Reserve Corps, Jan. 28, 1865.

Nichols, Daniel. Residence, Coventry. 21. S. Farmer. Enlisted Aug. 11, 1862. Mustered in Sept. 6, 1862. Transferred to Co. I, Feb. 1, 1865.

Nolan, Patrick. Residence, Pawtucket. 33. M. Dryer. Enlisted Aug. 12, 1862. Mustered in Sept. 6, 1862. Transferred to Co. I, Feb. 1, 1865.

O'Brian, Timothy. Residence, Boston, MA. 30. S. Laborer. Enlisted Aug. 5, 1862. Mustered in Sept. 6, 1862. Captured at Cold Harbor, VA, June 7, 1864. Paroled at Vicksburg, MS, April 6, 1865. Mustered out June 9, 1865.

O'Neil, Patrick. Residence, Johnston. 23. S. Laborer. Enlisted Aug. 9, 1862. Mustered in Sept. 6, 1862. Transferred to Co. I, Feb. 1, 1865.

Paine, George C. Residence, Burrillville. 18. S. Farmer. Enlisted Aug. 6, 1862. Mustered in Sept. 6, 1862. Discharged for disability Feb. 26, 1863. Died Aug. 23, 1894. Interred at Woodlawn Cemetery, Kansas City, KS.

Pierce, Christopher R. Residence, Coventry. 29. M. Jack spinner. Enlisted Aug. 6, 1862. Mustered in Sept. 6, 1862. Wounded in action, shot in hip, at Fredericksburg, VA, Dec. 13, 1862. Died of dysentery in camp at Milldale, MS, July 9, 1863. Cenotaph at Knotty Oak Cemetery, Coventry, RI.

Raferty, Peter. Residence, Providence. 35. M. Laborer. Enlisted Aug. 8, 1862. Mustered in Sept. 6, 1862. Deserted in Baltimore, MD, Mar. 27, 1863.

Sherman, Daniel B. Residence, East Greenwich. 17. S. Farmer. Enlisted Aug. 8, 1862. Mustered in Sept. 6, 1862. Promoted to corporal.

Stanfield, William. Residence, Johnston. 44. M. Farmer. Enlisted Aug. 10, 1862. Mustered in Sept. 6, 1862. On extra duty in quartermaster's department from Nov. 1862, and so borne until Feb. 1863. Transferred to Co. I, Feb. 1, 1865.

Steere, Benoni. Residence, Burrillville. 33. M. Farmer. Enlisted Aug. 7, 1862. Mustered in Sept. 6, 1862. Died of dysentery at Falmouth, VA, Dec. 23, 1862.

Sunderland, George B. Residence, Exeter. 35. S. Laborer. Enlisted Aug. 9, 1862. Mustered in Sept. 6, 1862. Wounded in action, shot in head and shoulder, at Petersburg, VA, June 22, 1864. Transferred to Co. I, Feb. 1, 1865.

Taylor, Richard Edwin. Residence, Scituate. 18. S. Farmer. Enlisted July 30, 1862. Mustered in Sept. 6, 1862. Transferred to Co. I, Feb. 1, 1865.

Webb, William W. Residence, Providence. 30. M. Clerk. Enlisted Aug. 11, 1862. Mustered in Sept. 6, 1862. Promoted to second lieutenant Co. B, Mar. 1, 1863.

Whipple, Albert H. Residence, Providence. 36. S. Harness maker. Enlisted Aug. 6, 1862. Mustered in Sept. 6, 1862. Wounded in action, shot in leg, at Bethesda Church, VA, June 3, 1864. Transferred to Co. I, Feb. 1, 1865.

Whipple, Olney. Residence, Burrillville. 22. S. Farmer. Enlisted Aug. 7, 1862. Mustered in Sept. 6, 1862. Died of Yazoo Fever in hospital at Nicholasville, KY, Sept. 10, 1863.

Whitman, Squire F. Residence, Coventry. 39. M. Weaver. Enlisted Aug. 6, 1862. Mustered in Sept. 6, 1862. Transferred to Co. I, Feb. 1, 1865.

Wilcox, Joseph P. Residence, Little Compton. 37. M. Weaver. Enlisted Aug. 11, 1862. Mustered in Sept. 6, 1862. Discharged for disability Mar. 11, 1863. Died July 14, 1907. Interred at Westfield Cemetery, Danielson, CT.

Wood, Frederick. Residence, North Providence. Enlisted Aug. 5, 1862. Mustered in Sept. 10, 1862. Transferred to Co. I, Feb. 1, 1865.

COMPANY E

Captains

Bates, Gustavus D. Promoted from first lieutenant Co. K, July 25, 1864. Mustered out Nov. 2, 1864. Died July 24, 1911. Interred at Grove Street Cemetery, Putnam, CT.

Daniels, Percy. Promoted from first lieutenant Co. E, Mar. 1, 1863. Promoted to lieutenant colonel June 29, 1864.

Hunt, Edwin L. Transferred from Co. G, Oct. 21, 1864. Wounded in action at Petersburg, VA, April 2, 1865. Borne as absent sick until May 15, 1865, when he reported for duty. Mustered out July 26, 1865.

Tobey, Thomas F. Residence, Providence. 21. S. Lawyer. Commissioned Sept. 4, 1862. Mustered in Sept. 6, 1862. Wounded in action, shell fragment to left hand, at Fredericksburg, VA, Dec. 13, 1862. Promoted to major Jan. 7, 1863.

First Lieutenants

Daniels, Percy. Residence, Woonsocket. 21. S. Engineer. Commissioned Sept. 4, 1862. Mustered in Sept. 6, 1862. Promoted to captain Co. E, Mar. 1, 1863.

Johnson, William H. Promoted from first sergeant Co. D, Oct. 21, 1864. Mustered out June 9, 1865. Died June 10, 1898. Interred at Cambridge, MA.

Peckham, Peleg E. Promoted from second lieutenant Co. E, Mar. 1, 1863. Promoted to captain Co. B, July 25, 1864.

Second Lieutenants

Bates, Gustavus D. Promoted from sergeant Co. E, Mar. 1, 1863. Promoted to first lieutenant Co. K May 1863.

Brownell, Dexter L. Promoted from sergeant Co. H, May 23, 1863. Resigned April 20, 1864. Died Jan. 7, 1920. Interred at Swan Point Cemetery, Providence, RI.

Peckham, Peleg E. Promoted from sergeant Co. A, Jan. 7, 1863. Promoted to first lieutenant Co. E Mar. 1, 1863.

Wilbur, George A. Residence, Woonsocket. 30. S. Lawyer. Commissioned Sept. 4, 1862. Mustered in Sept. 6, 1862. Wounded in action, shot in thigh, at Fredericksburg, VA, Dec. 13, 1862. Borne as absent on account of wounds until Feb. 1863. Commissioned first lieutenant Jan. 7, 1863, and transferred to Co. K, Jan. 13, 1863.

First Sergeants

Porter, Charles L. Promoted from corporal Oct. 2, 1862. Wounded in action, shot in leg, at Bethesda Church, VA, June 3, 1864. Mustered out June 9, 1865. Died May 12, 1900. Interred at East Thompson Cemetery, Thompson, CT.

Roberts, Henry. Residence, Keene, NH. 26. S. Fireman. Enlisted July 3, 1862. Mustered in Sept. 6, 1862. Rolls for Sept. and Oct. 1862, reports him absent with remark: "Claimed as a deserter from Fifth Connecticut Regiment, and transferred to said regiment by order of commanding officer Oct. 2, 1862." Died Feb. 24, 1885. Interred at Auburn, NY.

Sergeants

Bates, Gustavus D. Residence, Thompson, CT. 23. S. Teacher. Enlisted Sept. 6, 1862. Mustered in Sept. 6, 1862. Promoted to second lieutenant Co. D, Mar. 1, 1863.

Bisbee, William A. Residence, Cumberland. 30. M. Painter. Enlisted July 31, 1862. Mustered in Sept. 6, 1862. Mustered out

June 9, 1865. Died July 18, 1873. Interred at Oak Grove Cemetery, Fall River, MA.

Boyden, Decatur M. Promoted from corporal. Wounded in action, in side, at the Wilderness, VA, May 6, 1864. Transferred to the Veteran Reserve Corps Sept. 30, 1864.

Greene, Esek. Promoted from corporal. Borne as on detached service at Conscript Camp, New Haven, CT, from Sept. 1863. Mustered out at Providence, June 30, 1865. Died Feb. 7, 1901. Interred at Oak Grove Cemetery, Pawtucket, RI.

Higgins, Thomas J. Promoted and Transferred from Co. C, Feb 15, 1865. Mustered out June 9, 1865. Died Sept. 17, 1895. Interred at North Burial Ground, Providence, RI.

Horton, Alonzo. Residence, Cumberland. 27. M. Painter. Enlisted July 21, 1862. Mustered in Sept. 6, 1862. Discharged for disability at Pleasant Valley, MD, Oct. 25, 1862.

Perry, Alpheus S. Residence, Cumberland. 21. S. Painter. Enlisted Aug. 5, 1862. Mustered in Sept. 6, 1862. Discharged for disability at Washington, DC, Jan. 24, 1863. Died of typhoid at Pawtucket, RI, Feb. 16, 1863. Interred at Mineral Spring Cemetery, Pawtucket, RI.

Warfield, Aaron B. Promoted from corporal. Wounded in action at the North Anna River, VA, May 25, 1864. Discharged for disability at Portsmouth Grove, RI, June 3, 1865. Interred at Oak Grove Cemetery, Woonsocket, RI.

Whitcomb, Andrew J. Residence, Providence. 28. S. Painter. Enlisted July 31, 1862. Mustered in Sept. 6, 1862. Wounded in action at Poplar Spring Church, VA, Sept. 30, 1864. Borne as absent sick until June 1865. Mustered out July 17, 1865. Died April 6, 1906. Interred at Mount Caesar Cemetery, Swanzey, NH.

Corporals

Armstrong, Charles H. Residence, Cumberland. 39. M. Carder. Enlisted Aug. 9, 1862. Mustered in Sept. 6, 1862. Wounded in action, right foot amputated, at Fredericksburg, VA, Dec. 13, 1862. Discharged for disability May 4, 1863. Died Oct. 26, 1893. Interred at Mineral Springs Cemetery, Pawtucket, RI.

Arnold, Emory J. Promoted from private. Mustered out June 9, 1865. Died June 18, 1899. Interred at Union Cemetery, North Smithfield, RI.

Borden, Francis M. Residence, Cumberland. 22. M. Shoemaker. Enlisted July 29, 1863. Mustered in Sept. 6, 1862. Discharged for disability Mar. 2, 1863.

Boyden, Decatur M. Residence, Smithfield. 21. S. Laborer. Enlisted July 26, 1862. Mustered in Sept. 6, 1862. Wounded in action, shot in hand, at Fredericksburg, VA, Dec. 13, 1862. Sent to Washington and borne as absent sick until Feb. 1863. Promoted to sergeant.

Childs, Jonathan G. Promoted from private. Discharged for disability Mar. 2, 1863. Interred at Edgell Grove Cemetery, Framingham, MA.

Greene, Esek. Residence, Burrillville. 28. M. Spinner. Enlisted Aug. 8, 1862. Mustered in Sept. 6, 1862. Promoted to sergeant.

Howland, Andrew V. Residence, Burrillville. 18. S. Shoemaker. Enlisted Aug. 10, 1862. Mustered in Sept. 6, 1862. Left sick at Pleasant Valley, MD, Oct. 1862, and borne as absent sick until Jan. 1863. Mustered out June 9, 1865. Died Jan. 9, 1902. Interred at Porter Cemetery, Thompson, CT.

Porter, Charles L. Residence, Burrillville. 21. S. Shoemaker. Enlisted Aug. 7, 1862. Mustered in Sept. 6, 1862. Promoted to first sergeant Oct. 2, 1862. Served in Color Guard.

Proulx, John. Promoted from private. Mustered out June 9, 1865.

Sprague, Gilbert F. Promoted from private. Wounded in action at Spotsylvania Court House, VA, May 13, 1864. Mustered out June 9, 1865. Died 1917. Interred at Pine Grove Cemetery, Coventry, RI.

Warfield, Aaron B. Residence, Cumberland. 17. S. Clerk. Enlisted Aug. 5, 1862. Mustered in Sept. 6, 1862. Wounded in action, shot in right arm, at Fredericksburg, VA, Dec. 13, 1862. Sent to Washington and borne as absent sick until Feb.1863. Promoted to sergeant. Served in Color Guard.

Privates

Alexander, Hartford. Residence, Cumberland. 25. S. Farmer. Enlisted Aug. 2, 1862. Mustered in Sept. 6, 1862. Killed in action at Bethesda Church, VA, June 3, 1864. Interred at Cold Harbor National Cemetery. Grave 795.

Arnold, Daniel. Residence, Cumberland. 44. S. Spinner. Enlisted Aug. 6, 1862. Mustered in Sept. 6, 1862. Discharged for disability at Portsmouth Grove, RI, Jan. 4, 1864. Interred at Moshassuck Cemetery, Central Falls, RI.

Arnold, Emory J. Residence, Cumberland. 31. M. Butcher. Enlisted Aug. 21, 1862. Mustered in Sept. 6, 1862. Promoted to corporal.

Baker, William A. Residence, Coventry. 23. M. Laborer. Enlisted Aug. 14, 1862. Mustered in Sept. 6, 1862. Wounded in action, shot in knee, at Bethesda Church, VA, June 3, 1864. Mustered out June 9, 1865. Died Mar. 5, 1919. Interred at Yantic Cemetery, Norwich, CT.

Bates, George A. Residence, Scituate. 16. S. Farmer. Enlisted Aug. 14, 1862. Mustered in Sept. 6, 1862. Wounded in action, shot in head and arm, at Fredericksburg, VA, Dec. 13, 1862. Sent to hospital at Washington and borne as absent sick until Feb. 1863. Mustered out June 9, 1865. Died 1923. Interred at Cottrell Cemetery, Scituate, RI.

Blanchard, John E. Residence, Warwick. 18. S. Laborer. Enlisted Aug. 7, 1862. Mustered in Sept. 6, 1862. Temporarily detached to Battery D, First R. I. Light Artillery, from Jan. 15, 1863, until Feb. 1, 1865. Mustered out June 9, 1865. Died Aug. 20, 1897. Interred at Manchester Cemetery, Coventry, RI.

Boyle, Charles. Residence, Johnston. 23. S. Spinner. Enlisted Aug. 16, 1862. Mustered in Sept. 6, 1862. Mortally wounded in action at Fredericksburg, VA, Dec. 13, 1862. Died of wounds at Emory Hospital, Washington, DC, Feb. 1, 1863. Interred at Oak Grove Cemetery, Pawtucket, RI.

Boyle, William. Residence, Portsmouth. 21. S. Laborer. Enlisted Aug. 9, 1862. Mustered in Sept. 6, 1862. Wounded in action, shot in chest, at Fredericksburg, VA, Dec. 13, 1862. Wounded in action at Spotsylvania Court House, VA, May 18, 1864. Absent sick until May 19, 1865, when he was mustered out from Satterlee Hospital, Philadelphia, PA. Died April 11, 1916. Interred at Pocasset Cemetery, Cranston, RI.

Brennan, Thomas. Residence, Providence. 38. M. Gardiner. Enlisted and Mustered Aug. 17, 1864. Transferred to Co. D, June 9, 1865.

Briggs, Irvin D. Residence, Cumberland. 23. M. Machinist. Enlisted July 29, 1862. Mustered in Sept. 6, 1862. Wounded in action at Fredericksburg, VA, Dec. 13, 1862. Wounded in action, shot in side, at Jackson, MS, July 13, 1863. Mustered out June 9, 1865. Died Oct. 21, 1903. Interred at Old Town Cemetery, North Smithfield, RI.

Butler, Timothy. Residence, Providence. 19. S. Laborer. Enlisted Aug. 15, 1862. Mustered in Sept. 6, 1862. Captured June 6, 1864 at Cold Harbor, VA. Paroled at Vicksburg, MS, April 6, 1865. Mustered out June 30, 1865.

Cahoone, Sylvester. Residence, Exeter. 26. S. Farmer. Enlisted Aug. 12, 1862. Mustered in Sept. 6, 1862. Died of typhoid at Pleasant Valley, MD, Nov. 16, 1862.

Campbell, Peter A. Residence, Cumberland. 20. M. Spinner. Enlisted July 29, 1862. Mustered in Sept. 6, 1862. Discharged for disability at Alexandria, VA, Dec. 21, 1862. Died 1891. Interred at Allenstown Cemetery, Allenstown, NH.

Childs, Jonathan G. Residence, Cumberland. 23. S. Clerk. Enlisted Aug. 5, 1862. Mustered in Sept. 6, 1862. Promoted to corporal.

Cornell, William A. Residence, Cumberland. 33. M. Wheelwright. Enlisted Aug. 15, 1862. Mustered in Sept. 6, 1862. In quartermaster's department Mar. 1865. At division headquarters from April until June 1865. Mustered out June 9, 1865. Died Sept. 16, 1873. Interred at Oak Grove Cemetery, Pawtucket, RI.

Connelly, Simeon. Residence, Providence. Enlisted and Mustered April 6, 1865. Transferred to Co. D, June 9, 1865.

Darling, Patrick. Residence, Cumberland. 45. M. Laborer. Enlisted Aug. 13, 1862. Mustered in Sept. 6, 1862. Wounded in action, shot in right foot, at Fredericksburg, VA, Dec. 13, 1862. Sent to hospital and borne as absent sick until Feb. 3, 1863, when he was discharged for disability at Stanton Hospital, Washington, D. C.

Dempster, John. Residence, Providence. 30. M. Laborer. Enlisted Aug. 13, 1862. Mustered in Sept. 6, 1862. Killed in action at Fredericksburg, VA, Dec. 13, 1862.

Dexter, Alonzo. Residence, Cumberland. 20. S. Farmer. Enlisted Aug. 12, 1862. Mustered in Sept. 6, 1862. Borne as absent sick at Pleasant Valley, MD, from Oct. 27, 1862, until Feb. 1863. Wounded in action, shot through hips, at Bethesda Church, VA, June 3, 1864. Sent to hospital and borne as absent sick until June 3, 1865, when he was mustered out at Washington, D. C. Died Feb. 4, 1919. Interred at Walnut Hill Cemetery, Pawtucket, RI.

Egan, William J. Residence, Providence. 18. S. Farmer. Enlisted Aug. 12, 1862. Mustered in Sept. 6, 1862. Mustered out June 9, 1865. Interred at St. Francis Cemetery, Pawtucket, RI.

Essex, Richard. Residence, West Greenwich. 24. S. Farmer. Enlisted Aug. 12, 1862. Mustered in Sept. 6, 1862. Died of dysentery at Lexington, KY, Sept. 23, 1863. Interred at Camp Nelson National Cemetery. Section D, Grave 1263.

Gill, John. Residence, Providence. 29. M. Shoemaker. Enlisted and Mustered Oct. 6, 1864. Transferred to Co. D, June 9, 1865.

Gill, William. Residence, Cumberland. 45. M. Weaver. Enlisted July 29, 1862. Mustered in Sept. 6, 1862. Wounded in action at Fredericksburg, VA, Dec. 13, 1862, and borne as absent sick until Feb. 1863. Mustered out June 9, 1865. Died Dec. 29, 1893. Interred at East Killingly Cemetery, Killingly, CT.

Grant, George S. Residence, Cumberland. 20. S. Jeweler. Enlisted Aug. 9, 1862. Mustered in Sept. 6, 1862. Discharged for disability for disability at Falmouth, VA, Dec. 9, 1862. Died of typhoid at Cumberland, RI, Jan. 16, 1863. Interred at Moshassuck Cemetery, Central Falls, RI.

Grant, Ira W. Residence, Cumberland. 18. S. Clerk. Enlisted Aug. 11, 1862. Mustered in Sept. 6, 1862. Killed in action at Bethesda Church, VA, June 3, 1864. Cenotaph at Moshassuck Cemetery, Central Falls, RI.

Grant, Samuel. Residence, East Providence. 20. S. Clerk. Enlisted Aug. 1, 1862. Mustered in Sept. 6, 1862. Discharged for disability from hospital Mar. 17, 1863.

Greene, Edward H. Residence, Cranston. 20. S. Clerk. Enlisted July 7, 1862. Mustered in Sept. 6, 1862. Transferred to Veteran Reserve Corps Sept. 12, 1863. Died Aug. 11, 1874. Interred at Quaker Burial Ground, Providence, RI.

Greene, Thomas W. Residence, West Greenwich. 17. S. Laborer. Enlisted Aug. 12, 1862. Mustered in Sept. 6, 1862. Mortally wounded in action, shot in hip, at Bethesda Church, VA, June 3, 1864. Sent to hospital and borne as absent sick until Jan. 28, 1865, when he was discharged for disability from Portsmouth Grove, RI.

Died of wounds at West Greenwich, RI, April 6, 1865. Interred at Wanton Greene Lot, Coventry Cemetery 128, Coventry, RI.

Greene, Wilbur T. Residence, Coventry. 34. M. Carpenter. Enlisted Aug. 14, 1862. Mustered in Sept. 6, 1862. Mustered out June 9, 1865. Died Aug. 15, 1896. Interred at Hopkins Hollow Cemetery, Coventry, RI.

Grinnell, John W. Residence, Burrillville. 23. S. Farmer. Enlisted Aug. 7, 1862. Mustered in Sept. 6, 1862. Transferred to Veteran Reserve Corps Oct. 31, 1863. Died Oct. 26, 1884. Interred at Westfield Cemetery, Danielson, CT.

Hagan, James. Residence, Warwick. 18. S. Carpenter. Enlisted Aug. 12, 1862. Mustered in Sept. 6, 1862. Mustered out June 9, 1865.

Hall, Caleb. Residence, West Greenwich. 36. S. Farmer. Enlisted Aug. 12, 1862. Mustered in Sept. 6, 1862. Wounded in action, shot in thigh, May 12, 1864, at Spotsylvania Court House, VA. Sent to hospital and borne as absent sick until June 17, 1865, when he was discharged for disability from Portsmouth Grove, RI. Died June 6, 1905. Interred at Swan Point Cemetery, Providence, RI.

Hartshorn, George H. Residence, North Providence. 18. S. Farmer. Enlisted Aug. 15, 1862. Mustered in Sept. 6, 1862. Mustered out June 9, 1865.

Holbrook, Joseph H. Residence, Glocester. 21. S. Mason. Enlisted Aug. 7, 1862. Mustered in Sept. 6, 1862. Died of heat stroke near Jackson, MS, July 21, 1863. Cenotaph at Acotes Hill Cemetery, Glocester, RI.

Horan, John. Residence, Cumberland. 24. M. Laborer. Enlisted Aug. 5, 1862. Mustered in Sept. 6, 1862. Mustered out June 9, 1865.

Johnson, William. Residence, West Greenwich. 24. M. Laborer. Enlisted Aug. 12, 1862. Mustered in Sept. 6, 1862. Wounded in action, left arm amputated, at Fredericksburg, VA, Dec. 13, 1862.

Discharged for disability at the Patent Office Hospital, Washington, DC, Jan. 15, 1863. Died Feb. 12, 1909. Interred at Greenwood Cemetery, Coventry, RI.

Joyeaux, Augustus. Residence, Providence. 27. M. Teacher. Enlisted Sept. 17, 1863. Mustered in Sept. 24, 1863. Wounded in action, shot in hand, at Spotsylvania Court House, VA, May 12, 1864. Mustered out May 29, 1865.

Keif, Patrick. Residence, Smithfield. 18. S. Laborer. Enlisted Aug. 8, 1862. Mustered in Sept. 6, 1862. Discharged for disability at Camp Dennison, OH, July 8, 1863.

Keith, George W. Residence, Providence. 18. S. Sailor. Enlisted Aug. 1, 1862. Mustered in Sept. 6, 1862. On duty in hospital department Jan. 1863. Wounded in action, shot in leg, at Spotsylvania Court House, VA, May 13, 1864. Mustered out June 9, 1865. Died Jan. 19, 1899. Interred at Arlington National Cemetery. Grave 13937.

Kelly, Patrick. Residence, Cumberland. 55. M. Tailor. Enlisted July 25, 1862. Mustered in Sept. 6, 1862. Killed in action at Fredericksburg, VA, Dec. 13, 1862.

Langley, James. Residence, Cumberland. 48. M. Overseer. Enlisted July 25, 1862. Mustered in Sept. 6, 1862. Discharged for disability at Falmouth, VA, Dec. 9, 1862. Interred at Moshassuck Cemetery, Central Falls, RI.

McCasline, Thomas. Residence, Pawtucket. 45. M. Farmer. Enlisted Aug. 14, 1862. Mustered in Sept. 6, 1862. Died of typhoid at Windmill Point, VA, Feb. 1, 1863. Interred at North Burial Ground, Providence, RI.

McLeod, Murdock. Residence, Providence. 38. M. Dresser. Enlisted Aug. 13, 1862. Mustered in Sept. 6, 1862. Transferred to Veteran Reserve Corps Oct. 17, 1862. Died 1910. Interred at South Burial Ground, Warren, RI.

McMullen, Patrick. Residence, Smithfield. 34. M. Spinner. Enlisted July 29, 1862. Mustered in Sept. 6, 1862. Discharged for disability at Alexandria, VA, Dec. 24, 1862.

Malone, John. Residence, Cumberland. 35. M. Laborer. Enlisted Aug. 13, 1862. Mustered in Sept. 6, 1862. Died of typhoid at Falmouth, VA, Dec. 16, 1862.

Maloy, Thomas. Residence, Tiverton. 22. S. Spinner. Enlisted Aug. 13, 1862. Mustered in Sept. 6, 1862. Killed in action at Fredericksburg, VA, Dec. 13, 1862.

Moore, John. Residence, Boston, MA. 26. S. Laborer. Enlisted Aug. 28, 1862. Mustered in Sept. 6, 1862. Wounded in action, shot in knee, at Petersburg, VA, June 26, 1864. Wounded in action at Poplar Spring Church, VA, Oct. 1, 1864. Mustered out June 9, 1865.

Murray, Adams. Residence, Smithfield. 39. M. Brush maker. Enlisted July 13, 1862. Mustered in Sept. 6, 1862. Deserted at Sandoral, IL, June 7, 1863.

Murray, Christopher. Residence, Cumberland. 31. S. Farmer. Enlisted Aug. 1, 1862. Mustered in Sept. 6, 1862. Mustered out at Lexington, KY, Aug. 14, 1865.

Murray, Patrick. Residence, Providence. 33. M. Shoemaker. Enlisted Aug. 13, 1862. Mustered in Sept. 6, 1862. Wounded in action, shot in thigh, at Fredericksburg, VA, Dec. 13, 1862. Discharged for disability at Georgetown Hospital, Mar. 4, 1863.

Pelan, Robert T. Residence, Providence. 18. S. Book agent. Enlisted Aug. 14, 1862. Mustered in Sept. 6, 1862. Mortally wounded in action at Fredericksburg, VA, Dec. 13, 1862. Died of wounds at Falmouth, VA, Dec. 15, 1862.

Perkins, Charles H. Residence, Cumberland. 21. S. Clerk. Enlisted Aug. 5, 1862. Mustered in Sept. 6, 1862. Absent sick at Pleasant Valley, MD, from Oct. 27, 1862, until Jan. 1863. Wounded in action, shot in arm, at Bethesda Church, VA, June 3, 1864.

Discharged for disability May 29, 1865. Died July 11, 1925. Interred at Union Cemetery, North Smithfield, RI.

Pierce, Henry F. Residence, Scituate. 27. M. Laborer. Enlisted July 7, 1862. Mustered in Sept. 6, 1862. Absent sick at Pleasant Valley from Oct. 27, 1862, until Feb. 1863. Wounded in action, shot in hand, at Spotsylvania Court House, VA, May 12, 1864. Mustered out at Satterlee General Hospital, Philadelphia, PA, May 19, 1865. Died 1923. Interred at Pocasset Cemetery, Cranston, RI.

Proulx, John. Residence, Cumberland. 25. M. Farmer. Enlisted Aug. 16, 1862. Mustered in Sept. 6, 1862. Promoted to corporal.

Rice, Stephen. Residence, Cumberland. 18. S. Laborer. Enlisted Aug. 9, 1862. Mustered in Sept. 6, 1862. Wounded in action, shot in thigh, at Bethesda Church, VA, June 3, 1864. Discharged for disability at Fort Sedgwick, VA, Jan. 4, 1865.

Richmond, Preston B. Residence, Little Compton. 30. M. Merchant. Enlisted Aug. 12, 1862. Mustered in Sept. 6, 1862. On detached duty as postmaster Jan. 1863. On detached duty as postmaster at corps headquarters Jan. 1865, and so borne until June 1865. Mustered out June 9, 1865. Died Sept. 12, 1883. Interred at Union Cemetery, Little Compton, RI.

Riley, Philip. Residence, Cumberland. 28. M. Laborer. Enlisted Aug. 4, 1862. Mustered in Sept. 6, 1862. Wounded in action, shot in left shoulder, at Fredericksburg, VA, Dec. 13, 1862. Wounded in action, shot in leg, at Spotsylvania Court House, VA, May 13, 1864. Sent to hospital at Philadelphia, PA. Discharged for disability April 8, 1865.

Sisson, Benjamin F. Residence, West Greenwich. 18. S. Farmer. Enlisted Aug. 12, 1862. Mustered in Sept. 6, 1862. Killed in action at Spotsylvania Court House, VA, May 12, 1864. Cenotaph at Knotty Oak Cemetery, Coventry, RI.

Slocum, Charles T. Residence, Smithfield. 23. M. Weaver. Enlisted Aug. 23, 1862. Mustered in Sept. 6, 1862. Absent sick at Pleasant Valley, MD, from Oct. 27, 1862, until Jan. 1863.

Discharged for disability at Convalescent Hospital, VA, Jan. 15, 1863.

Snow, Paul. Residence, Coventry. 25. M. Farmer. Enlisted Aug. 14, 1862. Mustered in Sept. 6, 1862. Wounded in action at Fredericksburg, VA, Dec. 13, 1862, and sent to hospital. Discharged for disability at Portsmouth Grove, RI, May 29, 1863. Died 1902. Interred at St. Mary's Cemetery, West Warwick, RI.

Sprague, Gilbert F. Residence, West Greenwich. 21. M. Laborer. Enlisted Aug. 13, 1862. Mustered in Sept. 6, 1862. Promoted to corporal.

Sprague, Henry C. Residence, West Greenwich. 18. S. Laborer. Enlisted Aug. 13, 1862. Mustered in Sept. 6, 1862. Mustered out June 9, 1865. Died Dec. 15, 1923. Interred at Pine Grove Cemetery, Hopkinton, RI.

Staples, Charles A. Residence, Cumberland. 17. S. Laborer. Enlisted Aug. 31, 1862. Mustered in Sept. 6, 1862. Wounded in action, shot in right leg, at Fredericksburg, VA, Dec. 13, 1862. Mustered out June 9, 1865. Died 1922. Interred at North Burial Ground, Providence, RI.

Staples, Henry N. Residence, Cumberland. 41. M. Laborer. Enlisted Aug. 4, 1862. Mustered in Sept. 6, 1862. Wounded in action, shot in right leg, at Fredericksburg, VA, Dec. 13, 1862. Transferred to Veteran Reserve Corps, Sept. 8, 1863. Died Oct. 19, 1917. Interred at Hampton National Cemetery. Section A, Grave 11130.

Steere, Edward F. Residence, Cumberland. 30. M. Carder. Enlisted Aug. 16, 1862. Mustered in Sept. 6, 1862. Discharged for disability at College General Hospital, Georgetown, DC, Dec. 16, 1862. Died April 14, 1905. Interred at Union Cemetery, North Providence, RI.

Taylor, Charles M. Residence, Cranston. 29. M. Mason. Enlisted Aug. 8, 1862. Mustered in Sept. 6, 1862. Discharged for disability at Providence, Mar. 12, 1863. Interred at Littleneck Cemetery, East Providence, RI.

Taylor, Joseph. Residence, Burrillville. 18. S. Laborer. Enlisted Aug. 11, 1862. Mustered in Sept. 6, 1862. Orderly at brigade headquarters Jan. 1863 to June 1865. Slightly wounded, stabbed in right arm with bayonet, Aug. 18, 1864 at the Weldon Railroad, VA. Medal of Honor recipient for heroism at Petersburg, VA, Aug. 18, 1864. Mustered out June 9, 1865. Died Feb. 24, 1914. Interred at Edson Cemetery, Lowell, MA.

Tillinghast, Ira A. Residence, Coventry. 41. M. Farmer. Enlisted Aug. 14, 1862. Mustered in Sept. 6, 1862. Discharged for disability at Falmouth, VA, Jan. 17, 1863. Died Dec. 21, 1883. Interred at Plain Meeting House Cemetery, West Greenwich, RI.

Trainor, Michael. Residence, Providence. Enlisted and Mustered Nov. 12, 1864. Transferred to Co. D, June 9, 1865.

Turner, Charles. Residence, Cumberland. 34. M. Shoemaker. Enlisted Aug. 15, 1862. Mustered in Sept. 6, 1862. Captured at Cold Harbor, VA, June 6, 1864. Died of dysentery at Andersonville Prison, GA, July 9, 1864. Interred at Andersonville National Cemetery. Grave 3075.

Wallace, Henry C. Residence, Warwick. 19. S. Sailor. Enlisted July 16, 1862. Mustered in Sept. 6, 1862. Deserted at Richmond, KY, May 5, 1863.

Waterman, John S. Residence, Pawtucket. 28. M. Laborer. Enlisted Aug. 13, 1862. Mustered in Sept.6. On extra duty in quartermaster's department from Nov. 1862, until Feb. 1863. Discharged for disability at Falmouth, VA, Feb. 1, 1863.

Weeks, Studley. Residence, North Providence. 49. M. Laborer. Enlisted Aug. 12, 1862. Mustered in Sept. 6, 1862. Wounded in action at Fredericksburg, VA, Dec. 13, 1862. Discharged for disability Feb. 27, 1863.

COMPANY F

Captains

Bennett, Lyman M. Residence, Coventry. 25. M. Clerk. Commissioned Sept. 4, 1862. Mustered in Sept. 6, 1862. Resigned Jan. 7, 1863. Died 1913. Interred at Manchester Cemetery, Coventry, RI.

Joyce, William H. Promoted from first lieutenant Co. D, Jan. 7, 1863. Brevet major for gallantry at Spotsylvania Court House, VA, May 18, 1864. Mustered out at Providence, RI, July 22, 1865. Died May 6, 1890. Interred at Riverside Cemetery, Pawtucket, RI.

First Lieutenants

Bolles, Albert A. Promoted from second lieutenant Co. F, July 31, 1864. Member of a general court-martial at division headquarters from Jan. to April 1865. Mortally wounded in action, shot in throat, at Petersburg, VA, April 2, 1865. Died of wounds April 9, 1865 at Petersburg, VA. Interred at Mineral Spring Cemetery, Pawtucket, RI.

Root, Bridgman C. Residence, Providence. 26. S. Manufacturer. Commissioned Sept. 4, 1862. Mustered in Sept. 6, 1862. Borne on detached service as aid-de-camp to Brigadier General Gabriel Paul from Sept. 19, 1862 until Dec. 8, 1862, when he resigned. Died April 14, 1873 at Washington, DC. Interred at Arlington National Cemetery.

Stone, George N. Promoted from second lieutenant Co. B, Jan. 1, 1863. Transferred to Co. H, Mar. 20, 1863.

Second Lieutenants

Bolles, Albert A. Promoted from sergeant Co. D, Mar. 1, 1863. Wounded in action, shot in foot, at Spotsylvania Court House, VA, May 18, 1864. Promoted to first lieutenant Co. F, July 31, 1864.

Kellen, Charles H. Posthumously promoted from first sergeant Co. F, Jan. 1, 1863. Interred at Swan Point Cemetery, Providence, RI.

First Sergeants

Burgess, William B. Promoted from sergeant Jan. 1, 1863. Mustered out June 9, 1865. Died July 25, 1883. Interred at Attleboro, MA.

Kellen, Charles H. Residence, Cumberland. 20. S. Teacher. Enlisted July 16, 1862. Mustered in Sept. 6, 1862. Mortally wounded in action, shot in leg and amputated, at Fredericksburg, VA, Dec. 13, 1862. Died of wounds at Carver General Hospital, Dec. 29, 1862. Posthumously promoted to second lieutenant Co. F, Jan. 1, 1863.

Sergeants

Burgess, William B. Promoted from private. Promoted to first sergeant Jan. 1, 1863.

Carr, Jesse. Promoted from corporal. Mustered out June 9, 1865. Died Sept. 16, 1921. Interred at Knotty Oak Cemetery, Coventry, RI.

Linton, Jonathan. Residence, Exeter. 34. M. Moulder. Enlisted Aug. 12, 1862. Mustered in Sept. 6, 1862. Wounded in action, shot in finger, at Bethesda Church, VA, June 3, 1864. Mustered out June 9, 1865. Died Nov. 16, 1882. Interred at Patoka Cemetery, Patoka, IL.

Maynard, Mander A. Residence, Smithfield. 31. M. Teacher. Enlisted Aug. 5, 1862. Mustered in Sept. 6, 1862. Mustered out June 9, 1865. Died January 13, 1913. Interred at Hope Cemetery, Worcester, MA.

Murphy, Daniel. Residence, Tiverton. 27. S. Boot maker. Enlisted Aug. 13, 1862. Mustered in Sept. 6, 1862. Deserted while on furlough at Washington, DC, Mar. 9, 1863.

Rowe, Joseph. Residence, Providence. 44. M. Agent. Enlisted July 31, 1862. Mustered in Sept. 6, 1862. Mustered out June 9, 1865. Died May 13, 1873. Interred at North Burial Ground, Providence, RI.

Rowley, John H. Promoted from corporal. Mustered out June 9, 1865. Died Feb. 14, 1916. Interred at Riverside Cemetery, Pawtucket, RI.

Smith, Albert M. Residence, Pawtucket. 34. M. Overseer. Enlisted July 29, 1862. Mustered in Sept. 6, 1862. Wounded in action at Fredericksburg, VA, Dec. 13, 1862. Discharged for disability at General Hospital, Philadelphia, PA, Feb. 4, 1863. Died June 17, 1897. Interred at Old Town Burial Ground, North Smithfield, RI.

Corporals

Bassett, Nathan S. Promoted from private. Wounded in action, shot in shoulder, at Spotsylvania Court House, VA, May 18, 1864. Transferred to Veteran Reserve Corps, Oct. 30, 1864. Died Jan. 27, 1900. Interred at Oak Grove Cemetery, Pawtucket, RI.

Briggs, Rowland B. Residence, Exeter. 23. S. Farmer. Enlisted Aug. 15, 1862. Mustered in Sept. 6, 1862. Died of typhoid at Mansion House Hospital, Alexandria, VA, Nov. 21, 1862. Interred at Wood River Cemetery, Richmond, RI.

Carr, Jesse Jr. Promoted from private. Promoted to sergeant.

Chase, Charles F. Promoted from private. Mustered out June 9, 1865. Died Aug. 6, 1911. Interred at Elm Grove Cemetery, North Kingstown, RI.

Denison, John H. Residence, North Providence. 31. M. Machinist. Enlisted Aug. 7, 1862. Mustered in Sept. 6, 1862. Discharged for disability at Emory Hospital, Washington, DC, April 2, 1863.

Devitt, John M. Residence, Tiverton. 25. S. Laborer. Enlisted Aug. 14, 1862. Mustered in Sept. 6, 1862. Mortally wounded in action at Bethesda Church, VA, June 3, 1864. Died of wounds at Harewood General Hospital, Washington, DC. July 8, 1864.

Lewis, Nathan B. Promoted from private. Mustered out June 9, 1865. Died April 10, 1925. Interred at Elm Grove Cemetery, North Kingstown, RI.

Noyes, Thomas E. Residence, Smithfield. 20. S. Farmer. Enlisted Aug. 9, 1862. Mustered in Sept. 6, 1862. Mustered out June 9, 1865. Died Jan. 8, 1931. Interred at Hope Cemetery, Worcester, RI.

Pearce, Thomas D. Promoted from private. Borne as absent sick at Portsmouth Grove, RI, from May 15, 1864, until mustered out June 15, 1865. Died Feb. 22, 1895. Interred at Elm Grove Cemetery, North Kingstown, RI.

Rowley, John H. Residence, North Providence. 21. S. Grocer. Enlisted Aug. 13, 1862. Mustered in Sept. 6, 1862. Wounded in action, shot in finger, at Cold Harbor, VA, June 1, 1864. Wounded in action at the Crater, Petersburg, VA, July 30, 1864. Promoted to sergeant.

Rhowerts, Charles. Promoted from private. Killed in action at Spotsylvania Court House, VA, May 18, 1864.

Smith, Albert L. Residence, Pawtucket. 40. M. Machinist. Enlisted Aug. 14, 1862. Mustered in Sept. 6, 1862. Wounded in action at Fredericksburg, VA, Dec. 13, 1862. Promoted to second lieutenant Co. I, Mar. 17, 1863.

Wilcox, John T. Promoted from private. Wounded in action, shot in chest at Bethesda Church, VA, June 3, 1864. Discharged for disability at Portsmouth Grove, RI, May 18, 1865. Died Mar. 18, 1921. Interred at Elm Grove Cemetery, North Kingstown, RI.

Musician

Rowe, James E. Residence, Providence. 18. S. Machinist. Enlisted Aug. 30, 1862. Mustered in Sept. 6, 1862. Mustered out June 9,

1865. Died July 8, 1916. Interred at Masonic Cemetery, Custer City, OK.

Wagoner

Curtis, Samuel. Residence, Smithfield. 41. M. Hostler. Enlisted Aug 11, 1862. Mustered in Sept. 6, 1862. Discharged for disability at Providence June 23, 1863. Died Feb. 23, 1893. Interred at Hopkins Mills Cemetery, Foster, RI.

Privates

Albro, Wanton L. Residence, Exeter. 18. S. Farmer. Enlisted Aug. 13, 1862. Mustered in Sept. 6, 1862. Mustered out June 9, 1865. Died Oct. 30, 1916. Interred at Plain Meeting House Cemetery, West Greenwich, RI.

Bassett, Nathan S. Residence, Pawtucket. 40. M. Operative. Enlisted Aug. 12, 1862. Mustered in Sept. 6, 1862. Promoted to corporal.

Bates, James W. Residence, Exeter. 36. S. Farmer. Enlisted Aug. 13, 1862. Mustered in Sept. 6, 1862. Wounded in action at Fredericksburg, VA, Dec. 13, 1862. Wounded in action, shot in foot, at Spotsylvania Court House, VA, May 14, 1864. Wounded in action at Cold Harbor, VA, June 8, 1864. Mustered out June 9, 1865. Died Jan. 22, 1901. Interred at Chestnut Hill Cemetery, Exeter, RI.

Battey, Thomas. Residence, Woonsocket. 28. M. Painter. Enlisted Aug. 14, 1862. Mustered in Sept. 6, 1862. Wounded in action, shot in side, at Fredericksburg, VA, Dec. 13, 1862. Discharged for disability at Providence, RI, Mar. 17, 1863.

Booth, William J. Residence, Providence. 39. M. Police officer. Enlisted Mar. 31, 1864. Mustered in April 18, 1864. Brigade pioneer from Mar. 1865, until July 1865. Transferred to Co. D, June 9, 1865.

Brown, Albert G. Residence, Exeter. 19. S. Farmer. Enlisted Aug. 14, 1862. Mustered in Sept. 6, 1862. Died of pneumonia at Newport News, VA, Feb. 27, 1863. Interred at Elm Grove Cemetery, North Kingstown, RI.

Brown, John F. Residence, Exeter. 21. S. Farmer. Enlisted Aug. 14, 1862. Mustered in Sept. 6, 1862. Died of typhoid at Mouth Pleasant Hospital, Washington, DC, Oct. 5, 1862. Interred at Elm Grove Cemetery, North Kingstown, RI.

Browning, Charles O. Residence, North Kingstown. 24. M. Blacksmith. Enlisted Aug. 13, 1862. Mustered in Sept. 6, 1862. Wounded in action, shot in chest, at Spotsylvania Court House, VA, May 18, 1864. Mustered out June 9, 1865. Died 1909. Interred at Evergreen Cemetery, Stonington, CT.

Burgess, William B. Residence, Providence. 22. S. Moulder. Enlisted Aug. 12, 1862. Mustered in Sept. 6, 1862. Promoted to sergeant.

Buxton, George. Residence, Smithfield. 38. M. Spinner. Enlisted Aug. 9, 1862. Mustered in Sept. 6, 1862. Mustered out June 9, 1865. Died May 28, 1907. Interred at Togus National Cemetery. Grave 2310.

Carr, Clark. Residence, Exeter. 18. S. Farmer. Enlisted Aug. 16, 1862. Mustered in Sept. 6, 1862. Discharged for disability Feb. 23, 1863. Died Aug. 4, 1899. Interred at Litchfield Cemetery, Hampton, CT.

Carr, Edward C. Residence, Smithfield. 21. S. Mason. Enlisted Smithfield Aug. 13, 1862. Mustered in Sept. 6, 1862. Wounded in action, shot in leg, at Spotsylvania Court House, VA, May 12, 1864. Mustered out June 9, 1865. Interred at Acotes Hill Cemetery, Glocester, RI.

Carr, Jesse, Jr. Residence, Exeter. 19. S. Farmer. Enlisted Aug. 13, 1862. Mustered in Sept 6, 1862. Promoted to corporal.

Chase, Charles F. Residence, North Kingstown. 18. S. Farmer. Enlisted Aug. 11, 1862. Mustered in Sept. 6, 1862. Promoted to corporal.

Chatter, Joseph. Residence, Exeter. 44. M. Machinist. Enlisted Sept. 10, 1862. Mustered in Sept. 10, 1862. Died of typhoid at Falmouth, VA, Dec. 3, 1862.

Clavin, Michael. Residence, Providence. Enlisted and Mustered Feb. 1, 1865. 35. M. Transferred to Co. D, June 9, 1865.

Davis, George W. Residence, Johnston. 30. S. Blacksmith. Enlisted Aug. 21, 1862. Mustered in Sept 6, 1862. Deserted Jan. 1865 at Alexandria, VA.

Deakin, Michael. Residence, North Kingstown. 23. S. Laborer. Enlisted Aug. 8, 1862. Mustered in Sept. 6, 1862. Deserted at York, PA, Mar. 27, 1863.

Dewey, Benjamin. Residence, South Kingstown. 28. M. Farmer. Enlisted Aug. 9, 1862. Mustered in Sept. 6, 1862. Transferred to the Veteran Reserve Corps Mar. 2, 1864. Interred at First Cemetery, East Greenwich, RI.

Eldred, John C. Residence, Johnston. 34. M. Blacksmith. Enlisted Aug. 8, 1862. Mustered in Sept. 6, 1862. On extra duty as brigade blacksmith from Nov. 1862 until Feb. 1863. Mustered out June 9, 1865. Died Jan. 20, 1892. Interred at Elm Grove Cemetery, North Kingstown, RI.

Fish, Ellery G. Residence, Tiverton. 21. S. Farmer. Enlisted Aug. 15, 1862. Mustered in Sept. 6, 1862. Deserted on the march to Winchester, KY, April 8, 1863.

Fisher, George. Residence, Smithfield. 41. M. Stage coach driver. Enlisted Aug. 8, 1862. Mustered in Sept. 6, 1862. Wounded in action at Fredericksburg, VA, Dec. 13, 1862. Discharged for disability Mar. 27, 1863. Died 1894. Interred at Swan Point Cemetery, Providence, RI.

Foster, Alonzo C. Residence, Pawtucket. 23. M. Laborer. Enlisted Aug. 12, 1862. Mustered in Sept. 6, 1862. Discharged for disability at Falmouth, VA, Dec. 9, 1862.

Franklin, Chester Lewis. Residence, Exeter. 25. S. Teacher. Enlisted Aug. 13, 1862. Mustered in Sept. 6, 1862. Wounded in action at Fredericksburg, VA, Dec. 13, 1862. Mortally wounded in action at the North Anna River, VA, May 25, 1864. Died of wounds at Port Royal, VA, May 28, 1864. Interred at Greenwood Cemetery, Coventry, RI.

Goddard, Joseph Jr. Residence, Cumberland. 18. S. Laborer. Enlisted Nov. 13, 1863. Mustered in Dec. 3, 1863. Captured at Petersburg, VA, June 16, 1864. Escaped and returned July 6, 1864. Transferred to Co. D, June 9, 1865.

Gorton, Elisha. Residence, Coventry. Enlisted and Mustered Feb. 9, 1865. Transferred to Co. D, June 9, 1865.

Hall, George C. Residence, North Kingstown. 20. M. Farmer. Enlisted Aug. 8, 1862. Mustered in Sept. 6, 1862. Mustered out June 9, 1865. Died Oct. 28, 1928. Interred at Hunt-Hall Cemetery, North Kingstown Cemetery 11, North Kingstown, RI.

Hollis, Eben. Residence, North Kingstown. 28. M. Laborer. Enlisted Aug. 8, 1862. Mustered in Sept. 6, 1862. Wounded in action, shot in side, at Petersburg, VA, June 17, 1864. Mustered out June 9, 1865. Died Dec. 5, 1902. Interred at Elm Grove Cemetery, North Kingstown, RI.

Hunt, Benjamin S. Residence, North Kingstown. 18. S. Farmer. Enlisted Aug. 8, 1862. Mustered in Sept. 6, 1862. Killed in action at Fredericksburg, VA, Dec. 13, 1862. Cenotaph at Hunt-Hall Cemetery, North Kingstown Cemetery 11, North Kingstown, RI.

Kennedy, Timothy. Residence, Smithfield. Enlisted and Mustered Feb. 14, 1865. Sent to hospital Mar. 1865, and borne as absent sick until June 1865. Mustered out at Washington, DC, June 6, 1865.

Kenyon, Albert D. Residence, South Kingstown. 19. S. Farmer. Enlisted Aug. 13, 1862. Mustered in Sept. 6, 1862. Mortally wounded in action at Fredericksburg, VA, Dec. 13, 1862. Died of wounds at Falmouth, VA, Feb. 27, 1863.

Kenyon, Waite R. Residence, Richmond. 24. M. Farmer. Enlisted Aug. 15, 1862. Mustered in Sept. 6, 1862. Mustered out June 9, 1865. Died July 30, 1893. Interred at Pocasset Cemetery, Cranston, RI.

Knight, Thomas. Residence, Pawtucket. 44. M. Mule spinner. Enlisted Aug. 11, 1862. Mustered in Sept. 6, 1862. Mortally wounded in action at Fredericksburg, VA, Dec. 13, 1862. Died of wounds at Falmouth, VA, Dec. 15, 1862. Interred at Mineral Spring Cemetery, Pawtucket, RI.

Lewis, Nathan B. Residence, Exeter. 20. S. Teacher. Enlisted Aug. 13, 1862. Mustered in Sept. 6, 1862. Promoted to corporal.

Luther, John W. Residence, Exeter. 41. M. Machinist. Enlisted Aug. 12, 1862. Mustered in Sept. 6, 1862. Wounded in action, shot in back, arm, and hip at Spotsylvania Court House, VA, May 12, 1864. Transferred to Veteran Reserve Corps, Jan. 28, 1865. Died Jan. 25, 1905. Interred at Oak Grove Cemetery, Pawtucket, RI.

McClain, Thomas. Residence, Westerly. 35. M. Gardiner. Enlisted Aug. 13, 1862. Mustered in Sept. 6, 1862. Discharged for disability at Providence Mar. 17, 1863.

McDavitt, John. Residence, Tiverton. 21. S. Farmer. Enlisted Aug. 14, 1862. Mustered in Sept. 6, 1862. Mortally wounded in action at Bethesda Church, VA, June 3, 1864. Died of wounds at Harewood General Hospital, Washington, DC, July 8, 1864. Interred at Arlington National Cemetery. Section 13, Grave 6464.

McIntyre, Andrew. Residence, Providence. 38. M. Operative. Enlisted Aug. 15, 1862. Mustered in Sept. 6, 1862. Mustered out June 9, 1865.

McKenna, Charles. Residence, Cranston. 40. M. Driller. Enlisted Aug. 16, 1862. Mustered in Sept. 6, 1862. Deserted at Pleasant Valley, MD, Oct. 15, 1862.

McKenna, Patrick. Residence, Westerly. 34. M. Laborer. Enlisted Aug. 6, 1862. Mustered in Sept. 6, 1862. Wounded in action, shot in cheek, at Fredericksburg, VA, Dec. 13, 1862. Transferred to Veteran Reserve Corps Sept. 1, 1863. Interred at St. Michael's Cemetery, Stonington, CT.

Matthewson, Nicholas W. Residence, West Greenwich. 27. M. Laborer. Enlisted Aug. 8, 1862. Mustered in Sept. 6, 1862. Killed in action at Fredericksburg, VA, Dec. 13, 1862. Cenotaph at Elm Grove Cemetery, North Kingstown, RI.

Money, Joseph Jr. Residence, Exeter. 19. S. Farmer. Enlisted Aug. 15, 1862. Mustered in Sept. 6, 1862. Mustered out June 9, 1865. Died Oct. 29, 1897. Interred at Knotty Oak Cemetery, Coventry, RI.

Pate, William. Residence, Mansfield, MA. 35. S. Jeweler. Enlisted Providence Sept. 2, 1862. Mustered in Sept. 6, 1862. Killed in action at Bethesda Church, VA, June 3, 1864. Interred at Cold Harbor National Cemetery. Grave 828.

Pearce, Thomas D. Residence, North Kingstown. 35. M. Carpenter. Enlisted Aug. 9, 1862. Mustered in Sept. 6, 1862. Promoted to corporal.

Perkins, Palmer G. Residence, Exeter. 31. M. Spinner. Enlisted Aug. 12, 1862. Mustered in Sept. 6, 1862. Killed in action at Bethesda Church, VA, June 3, 1864. Interred at Cold Harbor National Cemetery. Grave 823. Cenotaph at Perkins Lot, Exeter Cemetery 80, Exeter, RI.

Phillips, Ezekiel B. Residence, North Kingstown. 37. M. Operative. Enlisted Aug. 9, 1862. Mustered in Sept. 6, 1862. Died of typhoid at Falmouth, VA, Dec. 9, 1862. Interred at Pearce-Phillips Lot, North Kingstown Cemetery 15, North Kingstown, RI.

Pierce, George S. Residence, North Kingstown. 27. M. Farmer. Enlisted Aug. 9, 1862. Mustered in Sept. 6, 1862. Mustered out June 9, 1865. Interred at Common Burial Ground, Newport, RI.

Pierce, Horatio N. Residence, North Providence. 44. M. Printer. Enlisted Aug. 14, 1862. Mustered in Sept. 6, 1862. Died of typhoid at Falmouth, VA, Dec. 19, 1862.

Potter, Benjamin Jr. Residence, North Kingstown. 42. M. Carpenter. Enlisted Aug. 8, 1862. Mustered in Sept. 6, 1862. Discharged for disability at Camp Dennison, OH, Nov. 17, 1863. Interred at Elm Grove Cemetery, North Kingstown, RI.

Reed, Frank E. Residence, Warwick. 18. S. Farmer. Enlisted Aug. 14, 1862. Mustered in Sept. 6, 1862. Wounded in action, in mouth, at Fredericksburg, VA, Dec. 13, 1862. Borne as absent sick until Jan. 1863. Died of dysentery at Milldale, MS, July 30, 1863. Interred near Milldale, MS.

Rex, Henry. Residence, Pawtucket. 18. S. Laborer. Enlisted Aug. 13, 1862. Mustered in Sept. 6, 1862. Wounded in action at Bethesda Church, VA, June 3, 1864. Mustered out June 9, 1865.

Rhowerts, Charles. Residence, Providence. 28. S. Boot maker. Enlisted Aug. 13, 1862. Mustered in Sept. 6, 1862. Wounded in action at Fredericksburg, VA, Dec. 13, 1862. Promoted to corporal.

Robinson, Levi. Residence, Providence. Enlisted and Mustered Feb. 24, 1865. Borne as absent sick in General Hospital, Alexandria, from May 9, 1865, until June. Mustered out at Washington, DC, June 15, 1865.

Rose, George P. Residence, North Kingstown. Enlisted Aug. 8, 1862. Mustered in Sept. 6, 1862. Absent sick in Frederick, MD, from Oct. 3, 1862, until Feb. 1863. Died of dysentery at North Kingstown, RI, Sept. 16, 1864, while on furlough from United States Post Hospital, Fort Wool, NY Harbor. Interred at Elm Grove Cemetery, North Kingstown, RI.

Rowley, Robert. Residence, North Providence. 28. M. Nail maker. Enlisted Aug. 14, 1862. Mustered in Sept. 6, 1862. Mustered out June 9, 1865. Interred at Riverside Cemetery, Pawtucket, RI.

Russell, William H. Residence, Tiverton. 22. M. Farmer. Enlisted Aug. 15, 1862. Mustered in Sept. 6, 1862. Wounded in action at Fredericksburg, VA, Dec. 13, 1862. Transferred to Veteran Reserve Corps, Oct. 17, 1863.

Sanderson, William. Residence, North Providence. 28. M. Operative. Enlisted Aug. 6, 1862. Mustered in Sept. 6, 1862. Discharged for disability at Portsmouth Grove, RI, Jan. 19, 1864.

Scully, Timothy. Residence, Coventry. 28. M. Spinner. Enlisted Aug. 18, 1862. Mustered in Sept. 6, 1862. Deserted while the regiment was passing through Baltimore, MD, Mar. 27, 1863.

Spencer, John. Residence, Exeter. 37. S. Farmer. Enlisted Aug. 13, 1862. Mustered in Sept. 6, 1862. Died of dysentery at Camp Dennison Hospital, OH, Sept. 14, 1863. Interred at Waldschmidt Cemetery, Camp Dennison, OH. Cenotaph at Elm Grove Cemetery, North Kingstown, RI.

Straight, Potter P. Residence, Exeter. 30. S. Farmer. Enlisted Aug. 15, 1862. Mustered in Sept. 6, 1862. Mortally wounded in action at Bethesda Church, VA, June 3, 1864. Died of wounds at Armory Square Hospital, Washington, DC, June 16, 1864. Interred at Ezekiel Austin Lot, Exeter Cemetery 58, Exeter, RI.

Sullivan, Timothy. Residence, Providence. Enlisted and Mustered Feb. 13, 1865. Discharged for disability at Alexandria, VA, June 24, 1865.

Taylor, John H. Residence, Warwick. 36. M. Machinist. Enlisted Aug. 8, 1862. Mustered in Sept. 6, 1862. On detached service in ambulance corps from Oct. 31, 1862, until Feb. 1863. Mustered out June 9, 1865. Died Oct. 11, 1900. Interred at Elm Grove Cemetery, North Kingstown, RI.

Tourgee, Charles S. Residence, North Kingstown. 37. M. Farmer. Enlisted Aug. 11, 1862. Mustered in Sept. 6, 1862. Absent sick at Alexandria, from May 11, 1864, until May 29, 1865, when he was mustered out at Washington, D. C. Died April 10, 1904. Interred at Elm Grove Cemetery, North Kingstown, RI.

Welden, William. Residence, Providence. 19. S. Laborer. Enlisted Sept. 7, 1863. Wounded in action, shot in face, at Cold Harbor, VA, June 2, 1864. Wounded in action, shot in arm, at Poplar Spring Church, VA, Oct. 1, 1864. Mustered out July 13, 1865.

Whitman, Thomas R. Residence, Providence. 14. Enlisted and Mustered Feb. 17, 1865. Transferred to Co. D, June 9, 1865.

Wilcox, John T. Residence, North Kingstown. 17. S. Teacher. Enlisted Aug. 12, 1862. Mustered in Sept. 6, 1862. Promoted to corporal.

COMPANY G

Captains

Greene, Thomas. Promoted from first lieutenant Co. G, Mar. 1, 1863. Resigned April 1, 1864. Died July 23, 1896. Interred at North Burial Ground, Providence, RI.

Hunt, Edwin L. Promoted from first lieutenant Co. I, May 3, 1864, but not mustered until Nov. 1, 1864. Transferred to Co. E, Oct. 21, 1864.

Rodman, Rowland G. Residence, South Kingstown. 34. M. Manufacturer. Commissioned Sept. 4, 1862. Mustered in Sept. 6, 1862. Wounded in action, shot in chest, at Fredericksburg, VA, Dec. 13, 1862. Discharged for disability Feb. 27, 1863. Died Mar. 10, 1901. Interred in I.P. Rodman Lot, South Kingstown Cemetery 30, South Kingstown, RI.

First Lieutenants

Channell, Alfred M. Residence, Providence. 30. S. Moulder. Commissioned Sept. 4, 1862. Mustered in Sept. 6, 1862. Promoted to captain Co. D, Oct. 24, 1862.

Greene, Thomas. Residence, South Kingstown. 49. M. Manufacturer. Commissioned and Mustered Oct. 25, 1862. Promoted to captain Co. G, Mar. 1, 1863.

Morse, Ephraim C. Promoted from second lieutenant Co. G, July 25, 1864. Promoted to regimental quartermaster Jan. 11, 1865.

Weigand, Frederick Promoted from second lieutenant Co. G, Mar. 1, 1863. Wounded in action, shot in hand, at Spotsylvania Court House, VA, May 12, 1864. Transferred to Co. B, July 31, 1864.

Second Lieutenants

Allen, Edward T. Residence, South Kingstown. 23. S. Clerk. Commissioned Sept. 4, 1862. Mustered in Sept. 6, 1862. Borne as absent with leave from Nov. 12, 1862, until Dec. 13, 1862. Promoted to first lieutenant Co. A, Jan. 7, 1863.

Morse, Ephraim C. Promoted from sergeant Co. I, April 15, 1863. Wounded in action, shot in face, at Spotsylvania Court House, VA, May 18, 1864. Promoted to first lieutenant Co. G, July 25, 1864.

Weigand, Frederick. Promoted from sergeant Co. K Jan. 7, 1863. Promoted to first lieutenant Co. G, Mar. 1, 1863.

First Sergeants

Chappell, Winfield S. Promoted from private Aug. 7, 1863. Commissioned second lieutenant May 5, 1864, but never mustered as such. Commissioned first lieutenant Co. B, Oct. 21, 1864.

Hull, John K. Promoted from sergeant Dec. 31, 1862. Killed in action at Jackson, MS, July 13, 1863. Interred near Jackson, MS. Cenotaph at Riverside Cemetery, South Kingstown, RI.

Sweatt, Joseph S. Residence, Boscawen, NH. 19. S. Clerk. Enlisted Aug. 18, 1862. Mustered in Sept. 6, 1862. Mortally wounded in action at Fredericksburg, VA, Dec. 13, 1862. Sent to Boscawen, NH to recover. Died of wounds and typhoid at Boscawen, NH, Mar. 6, 1863. Interred at Penacook Cemetery, Boscawen, NH.

Sergeants

Blanchard, Isaac. Residence, East Greenwich. 33. M. Fisherman. Enlisted Aug. 15, 1862. Mustered in Sept. 6, 1862. Transferred to Co. H, Feb. 1, 1865.

Donahue, Matthew. Promoted from corporal. Transferred to Co. H, Feb. 1, 1865.

Hull, John K. Residence, South Kingstown. 21. S. Teacher. Enlisted Aug. 9, 1862. Mustered in Sept. 6, 1862. Promoted to first sergeant Dec. 31, 1862.

Knowles, Charles A. Residence, South Kingstown. 36. M. Blacksmith. Enlisted Aug. 9, 1862. Mustered in Sept. 6, 1862. Killed in action at Fredericksburg, VA, Dec. 13, 1862. Cenotaph in James Knowles Lot, South Kingstown Cemetery 32, South Kingstown, RI.

Richter, Henry M. Promoted from private. Transferred to Co. K, Sept. 1, 1864.

Webster, John W. Residence, South Kingstown. 24. M. Blacksmith. Enlisted Aug. 9, 1862. Mustered in Sept. 6, 1862. Transferred to Co. H, Feb. 1, 1865.

Whitford, John R. Residence, South Kingstown. 25. M. Blacksmith. Enlisted Aug. 4, 1862. Mustered in Sept. 6, 1862. Transferred to Veteran Reserve Corps, Sept. 10, 1864. Died Jan. 25, 1916. Interred at Riverside Cemetery, South Kingstown, RI.

Corporals

Donahue, Mathew. Residence, South Kingstown. 23. M. Farmer. Enlisted Aug. 11, 1862. Mustered in Sept. 6, 1862. Promoted to sergeant.

Keaton, Daniel R. Residence, South Kingstown. 25. M. Laborer. Enlisted Aug. 20, 1862. Mustered in Sept. 6, 1862. Wounded in action, shot in shoulder, at the Crater, Petersburg, VA, July 30, 1864. Transferred to Co. H, Feb. 1, 1865.

Potter, Joseph J. Residence, Richmond. 23. M. Carpenter. Enlisted Aug. 11, 1862. Mustered in Sept. 6, 1862. Wounded in action, finger amputated, at Fredericksburg, VA, Dec. 13, 1862, and sent to Portsmouth Grove, RI. Discharged for disability at Providence, May 28, 1863. Interred at White Brook Cemetery, Richmond, RI.

Open, Manuel. Residence, South Kingstown. 29. M. Weaver. Enlisted Aug. 9, 1862. Mustered in Sept. 6, 1862. Wounded in action, shot in back, at Fredericksburg, VA, Dec. 13, 1862. Killed in action at Spotsylvania Court House, VA, May 18, 1864. Served in Color Guard. Interred at Fredericksburg National Cemetery. Grave 1143.

Quinlan, William S. Residence, South Kingstown. 20. S. Laborer. Enlisted Aug. 8, 1862. Mustered in Sept. 6, 1862. Transferred to Co. H, Feb. 1, 1865. Served in Color Guard.

Snow, Samuel J. Promoted from private. Died of dysentery in hospital at Covington, KY, May 1, 1863. Interred at Covington National Cemetery. Section G, Grave 1213. Cenotaph at Snow Lot, Exeter Cemetery 76, Exeter, RI.

Wilson, Benjamin A. Residence, South Kingstown. 30. M. Carpenter. Enlisted Aug. 20, 1862. Mustered in Sept. 6, 1862. Wounded in action, shot in back, at Fredericksburg, VA, Dec. 13, 1862. Wounded in action, shot in leg, at Spotsylvania Court House, VA, May 12, 1864. Transferred to Co. C, Feb. 1, 1865.

Musicians

Carpenter, James. Residence, South Kingstown. 19. S. Laborer. Enlisted Aug. 5, 1862. Mustered in Sept. 6, 1862. Promoted to principal musician Dec. 15, 1864, and transferred to non-commissioned staff Dec. 18, 1864.

Holland, Franklin D. Residence, South Kingstown. 18. S. Laborer. Enlisted Aug. 9, 1862. Mustered in Sept. 6, 1862. Detached for service at Alexandria, VA, May 1864, and so borne until June 1865. Transferred to Co. H, Feb. 1, 1865.

Wagoner

Sherman, Carder H. Residence, South Kingstown. 35. M. Jack spinner. Enlisted Aug. 9, 1862. Mustered in Sept. 6, 1862. On detached duty in quartermaster's department Nov. 1862. Wounded in action, shot in chest, at Petersburg, VA, June 17, 1864. Transferred to Co. I, Feb. 1, 1865.

Privates

Austin, Wanton G. Residence, South Kingstown. 26. S. Farmer. Enlisted Aug. 6, 1862. Mustered in Sept. 6, 1862. Died of Yazoo Fever onboard the *David Tatum,* Aug. 10, 1863. Cenotaph at Beriah B. Gardner Lot, South Kingstown Cemetery 13, South Kingstown, RI.

Baacke, George E. Residence, Providence. 38. M. Jeweler. Enlisted Aug. 22, 1863. Mustered in Sept. 22, 1863. Wounded in action, shot in finger, at Petersburg, VA, June 29, 1864. Transferred to Co. H, Feb. 1, 1865.

Bacon, James H. Residence, South Kingstown. 19. S. Farmer. Enlisted Aug. 9, 1862. Mustered in Sept. 6, 1862. Died of typhoid at Falmouth, VA, Jan. 24, 1863. Interred at Oak Dell Cemetery, South Kingstown, RI.

Barber, Gilbert M. Residence, Hopkinton. 48. S. Farmer. Enlisted Aug. 12, 1862. Mustered in Sept. 6, 1862. Wounded in action, shot in back, at Fredericksburg, VA, Dec. 13, 1862. Transferred to Co. H, Feb.1, 1865.

Barber, Jesse N. Residence, Hopkinton. 33. S. Carpenter. Enlisted Aug. 9, 1862. Mustered in Sept. 6, 1862. Killed in action at Fredericksburg, VA, Dec. 13, 1862. Cenotaph at Wood River Cemetery, Richmond, RI.

Barber, Israel A. Residence, Hopkinton. 19. S. Farmer. Enlisted Aug. 8, 1862. Mustered in Sept. 6, 1862. Died of Yazoo Fever onboard *David Tatum*, Mississippi River, Aug. 5, 1863. Cenotaph at Usquepaugh Cemetery, South Kingstown, RI.

Billington, Daniel R. Transferred from Co. B, Nov. 20, 1863. Wounded in action, shot in leg, at Petersburg, VA, June 17, 1864. Transferred to Co. H, Feb. 1, 1865.

Billington, Robertson G. Residence, South Kingstown. 25. M. Fisherman. Enlisted Aug. 9, 1862. Mustered in Sept. 6, 1862. Discharged for disability at Newport News, VA, Mar. 2, 1863. Died Mar. 10, 1896. Interred at Riverside Cemetery, South Kingstown, RI.

Blanchard, Ephraim A. Residence, South Kingstown. 37. M. Tailor. Enlisted Aug. 8, 1862. Mustered in Sept. 6, 1862. Transferred to the Veteran Reserve Corps, Sept. 16, 1863. Died Dec. 14, 1886. Interred at Riverside Cemetery, South Kingstown, RI.

Bollig, John N. Residence, Providence. 29. M. Mechanic. Enlisted Aug. 22, 1862. Mustered in Sept. 6, 1862. Wounded in action, shot in leg, at Fredericksburg, VA, Dec. 13, 1862. Sent to hospital and borne as absent sick until Feb. 1863. Transferred to the Veteran Reserve Corps, Sept. 16, 1863. Died of disease contracted in the service at Providence, RI, Aug. 29, 1865.

Boss, Joseph A. Residence, South Kingstown. 18. M. Farmer. Enlisted Aug. 12, 1862. Mustered in Sept. 6, 1862. Detached to Battery D, First RI Light Artillery from Jan. 1863 until Dec. 10, 1864, when he was returned to the 7th RI Vols. Transferred to Co. H, Feb. 1, 1865.

Borden, Thomas B. Residence, Hopkinton. 42. M. Blacksmith. Enlisted Aug. 12, 1862. Mustered in Sept. 6, 1862. Wounded in action at Fredericksburg, VA, Dec. 13, 1862. Discharged for disability at Washington, DC, Feb. 18, 1863. Died Aug. 10, 1885. Interred at Wood River Cemetery, Richmond, RI.

Brayman, Henry. Residence, South Kingstown. Enlisted Aug. 8, 1862. Mustered in Sept. 6, 1862. Wounded in action at Fredericksburg, VA, Dec. 13, 1862. Died of Yazoo Fever at Camp Nelson, KY, Sept. 14, 1863. Interred at Camp Nelson National Cemetery. Section D, Grave 1264.

Briggs, Lemuel A. Residence, South Kingstown. 18. S. Farmer. Enlisted Aug. 12, 1862. Mustered in Sept. 6, 1862. Transferred to Co. H, Feb. 1, 1865.

Briggs, Wanton S. Residence, Richmond. 23. S. Blacksmith. Enlisted Aug. 12, 1862. Mustered in Sept. 6, 1862. Transferred to Co. H, Feb. 1, 1865.

Browning, Orlando N. Residence, South Kingstown. 21. S. Farmer. Enlisted Aug. 22, 1862. Mustered in Sept. 6, 1862. Killed in action at Fredericksburg, VA, Dec. 13, 1862. Cenotaph at Browning Lot, South Kingstown Cemetery 63, South Kingstown, RI.

Burdick, Welcome C. Residence, Hopkinton. 57. M. Blacksmith. Enlisted Aug. 22, 1862. Mustered in Sept. 6, 1862. Mortally wounded in action at Fredericksburg, VA, Dec. 13, 1862. Died of wounds at Douglass General Hospital, Washington, DC, Dec. 26, 1862. Interred at Soldier's Home National Cemetery, Washington, DC. Section D, Grave 5209. Cenotaph at Riverbend Cemetery, Westerly, RI.

Cameron, Uz. Residence, South Kingstown. 51. M. Peddler. Enlisted Aug. 4, 1862. Mustered in Sept. 6, 1862. Drowned in the Mississippi River, June 9, 1863.

Card, Welcome H. Residence, South Kingstown. 23. S. Farmer. Enlisted Aug. 14, 1862. Mustered in Sept. 6, 1862. Wounded in action, shot through thighs, at Fredericksburg, VA, Dec. 13, 1862, and sent to hospital. Discharged for disability at Convalescent Camp, VA, July 2, 1863. Died 1914. Interred at Quaker Meeting Ground Cemetery, South Kingstown, RI.

Caswell, James D. Residence, South Kingstown. 20. M. Laborer. Enlisted Aug. 9, 1862. Mustered in Sept. 6, 1862. Wounded in action, shot in hand, at Spotsylvania Court House, VA, May 12, 1864. Sent to Mount Pleasant General Hospital, Washington, DC. Transferred to Co. H, Feb. 1, 1865.

Champlin, Charles E. Residence, South Kingstown. 20. S. Farmer. Enlisted Aug. 13, 1862. Mustered in Sept. 6, 1862. Died of dysentery at South Kingstown, RI July 21, 1863. Interred at Perryville Cemetery, South Kingstown, RI.

Chappell, Winfield S. Residence, South Kingstown. 20. S. Laborer. Enlisted Aug. 10, 1862. Mustered in Sept. 6, 1862. Promoted to first sergeant Aug. 7, 1863.

Clarke, Jonathan R. Residence, South Kingstown. 32. M. Farmer. Enlisted Aug. 9, 1862. Mustered in Sept. 6, 1862. Killed in action at Jackson, MS, July 13, 1863. Interred near Jackson, MS.

Connor, Peter. Residence, South Kingstown. Enlisted Aug. 9, 1862. Mustered in Sept. 6, 1862. Transferred to Co. H, Feb. 1, 1865.

Crandall, Courtland E. Residence, South Kingstown. 45. M. Mechanic. Enlisted Aug. 9, 1862. Mustered in Sept. 6, 1862. Discharged for disability at Falmouth, VA, Dec. 9, 1862. Died of typhoid at Norwich, CT, January 24, 1863.

Crandall, Elisha Kenyon. Residence, Richmond. 30. M. Weaver. Enlisted Aug. 15, 1862. Mustered in Sept. 6, 1862. Wounded in action, shot in thigh, at Fredericksburg, VA, Dec. 13, 1862. Discharged for disability Sept. 21, 1863. Died June 10, 1912. Interred at Wood River Cemetery, Richmond, RI.

Crandall, John H. Residence, South Kingstown. 22. S. Farmer. Enlisted Aug. 11, 1862. Mustered in Sept. 6, 1862. Discharged for disability at Pleasant Valley, MD, Oct. 25, 1862. Died Aug. 29, 1896. Interred at Clark Crandall Lot, South Kingstown Cemetery 66, South Kingstown, RI.

Crowley, Michael. Residence, South Kingstown. 21. S. Farmer. Enlisted Aug. 9, 1862. Mustered in Sept. 6, 1862. Wounded in action, shot in leg, at Spotsylvania Court House, VA, May 15, 1864. Admitted to Emory General Hospital, Washington DC, May 22, 1864. Transferred to the U.S. Navy Sept. 5, 1864. Interred at Portsmouth, NH.

Dexter, Henry R. Residence, Glocester. 29. S. Laborer. Enlisted Aug. 14, 1862. Mustered in Sept. 6, 1862. Transferred to Co. H, Feb. 1, 1865.

Donahue, Martin. Residence, South Kingstown. 22. S. Farmer. Enlisted Aug. 16, 1862. Mustered in Sept. 6, 1862. Transferred to Co. H, Feb. 1, 1865.

Eddy, John S. Residence, Cranston. 19. S. Laborer. Enlisted Aug. 20, 1862. Mustered in Sept. 6, 1862. Sick at Pleasant Valley, MD, from Oct. 27, 1862, until Feb. 1863. Killed in action June 8, 1864 at Cold Harbor, VA. Interred at Arlington National Cemetery. Section 27, Grave 473. Cenotaph at North Burial Ground, Providence, RI.

Effinger, Julius. Residence, Providence. 28. S. Tailor. Enlisted Aug. 22, 1863. Absent sick at City Point from Jan. until April 1865. Transferred to Co. H, Feb. 1, 1865.

Fessenden, Samuel. Residence, Pawtucket. 29. S. Clerk. Enlisted Aug. 18, 1862. Mustered in Sept. 6, 1862. Promoted to sergeant major and transferred to non-commissioned staff June 7, 1863.

Finley, William. Residence, South Kingstown. 22. S. Laborer. Enlisted Aug. 11, 1862. Mustered in Sept. 6, 1862. On extra duty in hospital department from Dec. 1862, until Feb. 1863. Died of dysentery at Covington, KY, Aug. 15, 1863. Interred at Covington National Cemetery. Section G, Grave 2014.

Gardiner, Charles W. Residence, South Kingstown. 19. S. Farmer. Enlisted Aug. 20, 1862. Mustered in Sept. 6, 1862. Died of typhoid at Marine Hospital, Cincinnati, OH, Aug. 24, 1863. Interred at Usquepaugh Cemetery, South Kingstown, RI.

Gallagher, Owen. Residence, South Kingstown. Enlisted Aug. 11, 1862. Mustered in Sept. 6, 1862. Killed in action at Fredericksburg, VA, Dec. 13, 1862.

Greene, Robert B. Residence, South Kingstown. 21. S. Farmer. Enlisted Aug. 9, 1862. Mustered in Sept. 6, 1862. Mortally wounded in action at Fredericksburg, VA, Dec. 13, 1862. Died of wounds at Washington, DC, Jan. 2, 1863. Interred at Soldier's Home National Cemetery, Washington, DC. Section H, Grave 3331.

Griffin, Joseph H. Residence, Charlestown. 20. S. Student. Enlisted Aug. 20, 1862. Mustered in Sept. 6, 1862. Transferred to Co. H, Feb 1, 1865.

Harvey, John, Jr. Residence, South Kingstown. 23. S. Farmer. Enlisted Aug. 8, 1862. Mustered in Sept. 6, 1862. Transferred to Co. H, Feb. 1, 1865.

Harvey, William B. Residence, South Kingstown. 30. S. Farmer. Enlisted Aug. 8, 1862. Mustered in Sept. 6, 1862. Transferred to Co. H, Feb. 1, 1865.

Healey, Horace D. Residence, South Kingstown. 21. S. Farmer. Enlisted Aug. 11, 1862. Mustered in Sept. 6, 1862. Died of dysentery at Milldale, MS, Aug. 2, 1863.

Holland, George A. Residence, South Kingstown. 28. M. Farmer. Enlisted Aug. 11, 1862. Mustered in Sept. 6, 1862. Teamster from Jan. 1865 to April 1865. Transferred to Co. H, Feb. 1, 1865.

Holland, George H. Residence, South Kingstown. 24. S. Clerk. Enlisted Aug. 9, 1862. Mustered in Sept. 6, 1862. Transferred to Co. H, Feb. 1, 1865.

Holland, Reuben, Jr. Residence, South Kingstown. 18. S. Carder. Enlisted Aug. 9, 1862. Mustered in Sept. 6, 1862. Wounded in action, shot in elbow, at Petersburg, VA, July 4, 1864. Transferred to Co. H, Feb. 1, 1865.

Holloway, Horace R. Residence, South Kingstown. 43. M. Mason. Enlisted Aug. 11, 1862. Mustered in Sept. 6, 1862. Wounded in action, shot in back, at Fredericksburg, VA, Dec. 13, 1862. Discharged for disability at Newport News, VA, Mar. 19, 1863. Interred at Holloway Lot, South Kingstown Cemetery 118, South Kingstown, RI.

Hopkins, George L. Residence, Providence. 31. M. Blacksmith. Enlisted Aug. 22, 1862. Mustered in Sept. 6, 1862. Transferred to Co. H, Feb. 1, 1865.

Jackson, Ambrose F. Residence, Providence. 29. M. Blacksmith. Enlisted Aug. 22, 1862. Mustered in Sept. 6, 1862. Wounded in action, shot in head, at Fredericksburg, VA, Dec. 13, 1862. Discharged for disability at Portsmouth Grove, RI, June 8, 1863.

Johnson, William H. Residence, South Kingstown. 26. M. Farmer. Enlisted Aug. 11, 1862. Mustered in Sept. 6, 1862. Killed in action at Petersburg, VA, June 22, 1864. Interred at Poplar Grove National Cemetery. Section C, Grave 2525.

Kenyon, Benjamin R. A. Residence, South Kingstown. 32. M. Farmer. Enlisted Aug. 14, 1862. Mustered in Sept. 6, 1862. Discharged for disability at Falmouth, VA, Dec. 9, 1862. Died 1917. Interred at White Brook Cemetery, Richmond, RI.

Kenyon, John C. Residence, South Kingstown. 30. M. Farmer. Enlisted Aug. 9, 1862. Mustered in Sept. 6, 1862. Killed in action at Fredericksburg, VA, Dec. 13, 1862.

Kenyon, Thomas G. Residence, South Kingstown. 25. M. Farmer. Enlisted Aug. 9, 1862. Mustered in Sept. 6, 1862. Died of typhoid at Emory Hospital, Washington, DC, Mar. 1, 1863. Interred at North Burial Ground, Providence, RI.

Kenyon, Welcome H. Residence, South Kingstown. 26. S. Farmer. Enlisted Aug. 22, 1862. Mustered in Sept. 6, 1862. Discharged for disability at Camp Banks, Alexandria, VA, Mar. 23, 1863. Died of tuberculosis contracted in the service Sept. 24, 1864 at Richmond, RI. Interred at Cross Mills Cemetery, Charlestown, RI.

Knowles, Alfred H. Residence, South Kingstown. 23. S. Farmer. Enlisted Aug. 12, 1862. Mustered in Sept. 6, 1862. Transferred to Co. H, Feb. 1, 1865.

Larkham, David L. Residence, Richmond. 37. M. Merchant. Enlisted Aug. 12, 1862. Mustered in Sept. 6, 1862. Transferred to the Veteran Reserve Corps Aug. 8, 1864. Died May 15, 1905. Interred at White Brook Cemetery, Richmond, RI.

Lawton, William O. Residence, South Kingstown. 29. M. Farmer. Enlisted Aug. 12, 1862. Mustered in Sept. 6, 1862. Wounded in action, shot in back, at Fredericksburg, VA, Dec. 13, 1862. Discharged from Portsmouth Grove, RI, May 21, 1864. Died Nov. 27, 1909. Interred at Elm Grove Cemetery, North Kingstown, RI.

May, Elisha G. Residence, South Kingstown. 18. S. Farmer. Enlisted Aug. 11, 1862. Mustered in Sept. 6, 1862. Died of dysentery at Nicholasville, KY, Aug. 29, 1863. Interred at Riverside Cemetery, South Kingstown, RI.

Moore, Winthrop A. Residence, Waltham, MA. 22. M. Merchant. Enlisted Aug. 22, 1862. Mustered in Sept. 6, 1862. Detached on duty as clerk in quartermaster's department from Nov. 1862 until Feb. 1863. Sent to Providence, RI, Dec. 16, 1862 with remains of Lt. Col. W.B. Sayles. Promoted to second lieutenant Mar. 1, 1863. Assigned to Co. A, Mar. 19, 1863.

Nicholas, Albert. Residence, South Kingstown. 21. S. Farmer. Enlisted Aug. 11, 1862. Mustered in Sept. 6, 1862. Borne as absent sick from Oct. 14, 1864, until April 1865. Transferred to Co. H, Feb. 1, 1865.

Northrup, William R. Residence, Providence. 23. M. Mechanic. Enlisted Aug. 22, 1862. Mustered in Sept. 6, 1862. Wounded in action, shot in mouth, at the North Anna River, VA, May 26, 1864. Wounded in action at Petersburg, VA, June 27, 1864. Transferred to Co. H, Feb. 1, 1865.

O'Neil, James. Residence, South Kingstown. 44. M. Laborer. Enlisted Aug. 8, 1862. Mustered in Sept. 6, 1862. Mortally wounded in action at Fredericksburg, VA, Dec. 13, 1862. Died of wounds at Falmouth, VA, Dec. 16, 1862. Interred at Fredericksburg National Cemetery. Grave 3506. Cenotaph at Oak Dell Cemetery, South Kingstown, RI.

Pollock, William J. Residence, South Kingstown. 18. S. Farmer. Enlisted Aug. 11, 1862. Mustered in Sept. 6, 1862. Killed in action at Fredericksburg, VA, Dec. 13, 1862.

Potter, Franklin H. Residence, South Kingstown. 23. S. Farmer. Enlisted Aug. 8, 1862. Mustered in Sept. 6, 1862. Detached to Battery D, 1st RI Light Artillery from Dec. 1862, until Dec. 10, 1864. Transferred to Co. H, Feb. 1, 1865.

Potter, Jared J. Residence, Richmond. 23. M. Farmer. Enlisted Aug. 11, 1862. Mustered in Sept. 6, 1862. Wounded in action at Jackson, MS, July 13, 1863. Transferred to Co. H, Feb. 1, 1865.

Rose, Robert N. Residence, South Kingstown. 31. S. Farmer. Enlisted Aug. 11, 1862. Mustered in Sept. 6, 1862. Mortally wounded in action at Fredericksburg, VA, Dec. 13, 1862. Died of wounds at regimental hospital, Falmouth, VA, Feb. 3, 1863. Interred at Robert Northrup Lot, North Kingstown Cemetery 88, North Kingstown, RI.

Richter, Henry M. Residence, Providence. 26. S. Clerk. Enlisted Oct. 10, 1863. Mustered in Oct. 14, 1863. Promoted to sergeant.

Sisson, Charles E. Residence, Richmond. 26. M. Farmer. Enlisted Aug. 13, 1862. Mustered in Sept. 6, 1862. Discharged for disability at Washington, DC, Feb. 18, 1863. Died Dec, 27, 1904. Interred at Riverbend Cemetery, Westerly, RI.

Sisson, Randall, Jr. Residence, Richmond. 23. M. Farmer. Enlisted Aug. 12, 1862. Mustered in Sept. 6, 1862. Wounded in action, shot in arm, at Jackson, MS, July 13, 1863. Died of Yazoo Fever at Cincinnati, OH, Aug. 28, 1863. Interred at Spring Hill National Cemetery. Grave 534.

Smith, Roderick D. Residence, South Kingstown. 21. S. Farmer. Enlisted Aug. 11, 1862. Mustered in Sept. 6, 1862. Wounded in action, shot in head at Spotsylvania Court House, VA, May 13, 1864. Killed in action at Spotsylvania Court House, VA, May 18, 1864. Interred at Fredericksburg National Cemetery. Grave 1142.

Smith, Daniel. Residence, Warwick. 18. S. Farmer. Enlisted Aug. 26, 1862. Mustered in Sept. 6, 1862. Killed in action at Fredericksburg, VA, Dec. 13, 1862.

Snow, Samuel J. Residence, Exeter. 34. M. Carpenter. Enlisted Aug. 9, 1862. Mustered in Sept. 6, 1862. Promoted to corporal.

Stone, Albert. Residence, South Kingstown. 44. M. Laborer. Enlisted Aug. 8, 1862. Mustered in Sept. 6, 1862. Transferred to Co. H, Feb. 1, 1865.

Sweet, Daniel. Residence, South Kingstown. 32. M. Enlisted Aug. 18, 1862. Mustered in Sept. 6, 1862. Brigade clerk in quartermaster's department Nov. 1862. On detached service at division headquarters from Dec. 9, 1862, until Jan. 1863. Discharged for disability at Cincinnati, OH, Mar. 2, 1863. Died 1912. Interred at Elm Grove Cemetery, North Kingstown, RI.

Tefft, Samuel S. Residence, South Kingstown. 23. S. Farmer. Enlisted Aug. 11, 1862. Mustered in Sept. 6, 1862. Wounded in action, shot in hand, at Spotsylvania Court House, VA, May 12, 1864. Borne as absent sick in hospital at Washington, DC, from Jan. 10, 1865, until May 10, 1865. Transferred to Co. H, Feb. 1, 1865.

Tourgee, William. Residence, South Kingstown. 27. S. Carder. Enlisted Aug. 11, 1862. Mustered in Sept. 6, 1862. Died of dysentery at Nicholasville, KY, Sept. 5, 1863. Interred at Gould-Tourgee Lot, South Kingstown Cemetery 35, South Kingstown, RI.

Underwood, Perry G. Residence, South Kingstown. 23. M. Farmer. Enlisted Aug. 11, 1862. Mustered in Sept. 6, 1862. Wounded in action, shot in left knee, at Fredericksburg, VA, Dec. 13, 1862. Died of dysentery at Cincinnati, OH, Aug. 23, 1863. Interred at Spring Grove National Cemetery. Grave 191. Cenotaph at Riverside Cemetery, South Kingstown, RI.

Veazy, Frederick. Residence, Providence. 27. M. Carpenter. Enlisted Aug. 23, 1862. Mustered in Sept. 6, 1862. Absent sick at Pleasant Valley, MD, from Oct. 27, 1862, until Feb. 1863. Discharged for disability at Convalescent Camp, VA, Feb. 13, 1863.

Wells, Benjamin E. Residence, South Kingstown. 23. S. Farmer. Enlisted Aug. 8, 1862. Mustered in Sept. 6, 1862. Wounded in action at Spotsylvania Court House, VA, May 18, 1864. Transferred to Co. H, Feb. 1, 1865.

COMPANY H

Captains

Remington, James H. Residence, Warwick. 23. S. Lawyer. Commissioned Sept. 4, 1862. Mustered in Sept. 6, 1862. Wounded in action, shot in jaw, at Fredericksburg, VA, Dec. 13, 1862. Brevet major for "gallant and meritorious services" at Fredericksburg, VA, Mar. 13, 1865. Transferred to Veteran Reserve Corps at Elmira, NY, May 2, 1863. Died Feb. 11, 1899. Interred at Greenwood Cemetery, Brooklyn, NY.

Stone, George N. Promoted from first lieutenant Co. H, Oct. 7, 1863. Mustered out July 27, 1865. Died Mar. 8, 1901. Interred at Spring Grove Cemetery, Cincinnati, OH.

First Lieutenants

Inman, George B. Residence, Burrillville. 19. S. Teacher. Commissioned Sept. 4, 1862. Mustered in Sept. 6, 1862. On detached service with ambulance corps from Oct. 15, until Dec. 1862. Resigned Dec. 28, 1862. Died April 12, 1920. Interred at Leavenworth National Cemetery. Fort Leavenworth, KS.

McKay, John Jr. Promoted from second lieutenant Co. H, Oct. 21, 1864. Transferred to Co. B, Feb. 1, 1865.

Phelps, James T. Transferred from Co. I, Feb. 1, 1865. Mustered out June 9, 1865. Died 1916. Interred at North Burial Ground, Bristol, RI.

Stone, George N. Transferred from Co. F, Mar. 20, 1863. Promoted to captain Co. H, Oct. 7, 1863.

Young, Henry. Promoted from second lieutenant Co. C, Aug. 14, 1863. Discharged for disability April 27, 1864. Died April 14, 1906. Interred at Swan Point Cemetery, Providence, RI.

Second Lieutenants

Jenks, Ethan Amos. Residence, Foster. 35. M. Farmer. Commissioned Aug. 5, 1862. Mustered in Sept. 6, 1862. Promoted to captain Co. I, Mar. 3, 1863.

McKay, John Jr. Promoted from sergeant Co. H, June 29, 1864. Wounded in action July 25, 1864 at Petersburg, VA. Promoted to first lieutenant Co. H, Oct. 21, 1864.

Young, Henry. Promoted from sergeant Co. D, Mar. 1, 1863. Commissioned first lieutenant Co. H, Aug. 14, 1863.

First Sergeants

Bezeley, Jeremiah P. Transferred from Co. B, Feb. 1, 1865. Mustered out June 9, 1865. Died April 26, 1910. Interred at Locust Grove Cemetery, Providence, RI.

Brownell, Dexter L. Residence, Smithfield. 30. S. Farmer. Enlisted Aug. 7, 1862. Mustered in Sept. 6, 1862. Promoted to second lieutenant Co. E, May 23, 1863.

Kilton, Winfield S. Promoted from sergeant May 23, 1863. Detached as chief clerk at brigade headquarters from Jan. 1865, until mustered out June 9, 1865. Died July 10, 1890. Interred at Knotty Oak Cemetery, Coventry, RI.

Sergeants

Donahue, Matthew. Transferred from Co. G, Feb. 1, 1865. Mustered out at Providence, June 15, 1865. Interred at St. Francis Cemetery, Pawtucket, RI.

Keegan, Thomas. Promoted from corporal July 1, 1863. Mustered out June 9, 1865. Died July 3, 1885. Interred at State Farm Cemetery, Cranston, RI.

Kilton, Winfield S. Promoted from corporal. Promoted to first sergeant May 23, 1863.

McKay, John Jr. Promoted from private. Wounded in action at Petersburg, VA, June 29, 1864. Promoted to second lieutenant Co. H, June 29, 1864.

Potter, Harrison W. Residence, Warwick. 23. S. Jeweler. Enlisted Aug. 15, 1862. Mustered in Sept. 6, 1862. Deserted at Baltimore, MD, Mar. 28, 1863.

Rice, Samuel E. Promoted from corporal. Killed in action at Spotsylvania Court House, VA, May 18, 1864. Interred at Fredericksburg National Cemetery. Grave 1161. Cenotaph in First Cemetery, East Greenwich, RI.

Spencer, James B. Residence, Warwick. 25. S. Teacher. Enlisted Aug. 12, 1862. Mustered in Sept. 6, 1862. Died of tuberculosis at Newport News Mar. 6, 1863. Interred at Wickes Lot, East Greenwich Cemetery 72, East Greenwich, RI.

Taylor, Wilfred P. Residence, Warwick. 23. S. Student. Enlisted Aug. 8, 1862. Mustered in Sept. 6, 1862. Wounded in action, shot in side, at Fredericksburg, VA, Dec. 13, 1862. Discharged for disability June 18, 1863. Died Sept. 4, 1897. Interred at Lowell Cemetery, Lowell, MA.

Trask, John F. Residence, Warwick. 28. S. Spinner. Enlisted Aug. 15, 1862. Mustered in Sept. 6, 1862. Wounded in action, shot in chest, at Fredericksburg, VA, Dec. 13, 1862. Transferred to Veteran Reserve Corps Oct. 31, 1863. Died Oct. 15, 1880. Interred at Brown Hill Cemetery, Indianapolis, IN.

Webster, John W. Transferred from Co. G, Feb. 1, 1865. Mustered out June 9, 1865. Died July 22, 1896. Interred at Riverside Cemetery, South Kingstown, RI.

Wood, William T. Promoted from corporal. Died of dysentery at Camp Nelson, KY, Sept. 10, 1863. Interred at Brayton Cemetery, Warwick, RI.

Corporals

Austin, Joseph. Promoted from private June 20, 1863. Wounded in action, shot in hand, at Petersburg, VA, June 16, 1864. Mustered out June 9, 1865. Interred at First Cemetery, East Greenwich, RI.

Blanchard, Isaac. Transferred from Co. G, Feb. 1, 1865. Mustered out June 9, 1865. Died May 19, 1906. Interred at Riverside Cemetery, South Kingstown, RI.

Brown, Samuel G. Promoted from private. Died of dysentery at Camp Dennison, OH, Aug. 26, 1863. Interred at Spring Grove National Cemetery. Grave 908.

Keegan, Thomas. Residence, Westerly. 37. M. Operative. Enlisted Aug. 15, 1862. Mustered in Sept. 6, 1862. Color guard from Nov. 1862, until Jan. 1863. Promoted to sergeant July 1, 1863. Served in Color Guard.

Kilton, Winfield S. Residence, West Greenwich. 19. S. Merchant. Enlisted Aug. 16, 1862. Mustered in Sept. 6, 1862. Promoted to sergeant.

Knowles, Alfred H. Transferred from Co. G, Feb. 1, 1865. Mustered out June 9, 1865. Died Feb. 12, 1911. Interred at Pocasset Cemetery, Cranston, RI.

Knowles, John F. Residence, East Greenwich. 26. S. Student. Enlisted Aug. 15, 1862. Mustered in Sept. 6, 1862. Clerk in brigade commissary department from Jan. 1865, until June. Mustered out June 9, 1865. Interred at White Brook Cemetery, Richmond, RI.

Lansing, John A. Residence, Warwick. 22. S. Student. Enlisted Aug. 12, 1862. Mustered in Sept. 6, 1862. Captured near White Sulphur Springs, VA, Nov. 5, 1862. Released Feb. 1863.

Discharged for disability at Camp Banks, Alexandria, VA, Aug. 14, 1863.

Ludowicy, John P. Residence, East Greenwich. 41. M. Merchant. Enlisted Aug. 15, 1862. Mustered in Sept. 6, 1862. Discharged for disability Jan. 1, 1863. Died April 17, 1888. Interred at North Burial Ground, Providence, RI.

Northrup, William R. Transferred from Co. G Feb. 1, 1865. Mustered out June 9, 1865. Died Dec. 3, 1901. Interred at Brayton Cemetery, Warwick, RI.

McFarland, Daniel. Promoted from private. Mustered out June 9, 1865. Died Mar. 12, 1908. Interred at St. Francis Cemetery, Pawtucket, RI

Quinlan, William S. Transferred from Co. G Feb. 1, 1865. Mustered out June 9, 1865.

Rice, Samuel E. Residence, East Greenwich. 19. S. Clerk. Enlisted Aug. 15, 1862. Mustered in Sept. 6, 1862. Promoted to sergeant.

Smith, Charles H. Residence, Warwick. 27. S. Farmer. Enlisted Aug. 15, 1862. Mustered in Sept. 6, 1862. Transferred to the Veteran Reserve Corps, Sept. 16, 1863.

Taylor, Isaac Y. Promoted from private June 20, 1863. Mustered out June 9, 1865. Died 1909. Interred at Elm Grove Cemetery, North Kingstown, RI.

Wood, William T. Residence, Warwick. 30. S. Fisherman. Enlisted Aug. 15, 1862. Mustered in Sept. 6, 1862. Promoted to sergeant.

Musicians

Holland, Francis B. Transferred from Co. G, Feb. 1, 1865. Mustered out at Washington, DC, June 5, 1865. Died Mar. 19, 1934. Interred at Oak Dell Cemetery, South Kingstown, RI.

Kenneth, William. Residence, Westerly. 31. M. Machinist. Enlisted Aug. 10, 1862. Mustered in Sept. 6, 1862. Mustered out June 9, 1865. Died May 23, 1891. Interred at Riverbend Cemetery, Westerly, RI.

Terry, Morris. Residence, North Providence. 19. S. Sailor. Enlisted Aug. 18, 1862. Mustered in Sept. 6, 1862. Detached to Battery D, 1st RI Light Artillery, from Jan. 15, 1863, until Dec. 10, 1864. Absent sick from Jan. 1865, until May. Mustered out May 12, 1865.

Privates

Albro, Edmund B. Residence, East Greenwich. 40. M. Laborer. Enlisted Aug. 11, 1862. Mustered in Sept. 6, 1862. Died of typhoid at regimental hospital at Falmouth, VA, Dec. 30, 1862. Interred at First Cemetery, East Greenwich, RI.

Aldrich, Charles W. Residence, Providence. 25. S. Mechanic. Enlisted Aug. 22, 1862. Mustered in Sept. 6, 1862. Wounded in action at Petersburg, VA, June 29, 1864. Absent sick at Washington, DC, from Jan. 1865, until May. Absent sick at Boston, May 1865. Discharged for disability at Boston, MA, July 12, 1865.

Aldrich, Nathan E. Residence, Newport. 24. S. Teamster. Enlisted Aug. 22, 1862. Mustered in Sept. 6, 1862. Absent sick Jan. 1865, and so borne until mustered out June 9, 1865.

Arnold, Daniel. Residence, East Greenwich. 37. M. Farmer. Enlisted Sept. 4, 1862. Mustered in Sept. 6, 1862. Discharged for disability Dec. 9, 1862. Died Dec. 31, 1891. Interred at Elm Grove Cemetery, North Kingstown, RI.

Arnold, Reuben. Residence, East Greenwich. 41. M. Teamster. Enlisted Aug. 11, 1862. Mustered in Sept. 6, 1862. Killed in action at Fredericksburg, VA, Dec. 13, 1862. Cenotaph at Vaughn-Arnold Lot, North Kingstown Cemetery 9, North Kingstown, RI.

Austin, Joseph. Residence, Warwick. 36. M. Machinist. Enlisted Aug. 15, 1862. Mustered in Sept. 6, 1862. Promoted to corporal June 20, 1863.

Ayers, Oliver L. Residence, Tiverton. 31. M. Farmer. Enlisted Aug. 14, 1862. Mustered in Sept. 6, 1862. Wounded in action, lost a leg, at Spotsylvania Court House, VA, May 18, 1864. Sent to hospital and borne as absent sick until June 23, 1865, when discharged for disability at Washington, DC.

Bailey, Martin. Residence, East Greenwich. 39. M. Farmer. Enlisted Aug. 11, 1862. Mustered in Sept. 6, 1862. Discharged for disability at post hospital, Lexington, KY, April 16, 1865.

Baggen, Bernard. Residence, Cranston. Enlisted and Mustered Mar. 13, 1865. Transferred to Co. D, June 9, 1865.

Baacke, George E. Transferred from Co. G, Feb. 1, 1865. Transferred to Co. D, June 9, 1865.

Barber, Gilbert M. Transferred from Co. G, Feb. 1, 1865. Mustered out June 9, 1865. Died May 11, 1907. Interred at Wood River Cemetery, Richmond, RI.

Battey, Joseph. Residence, Smithfield. 35. M. Weaver. Enlisted Aug. 10, 1862. Mustered in Sept. 6, 1862. Mustered out June 9, 1865. Died Aug. 20, 1906. Interred at Togus National Cemetery. Grave 2215.

Bicknell, Thomas W. Residence, East Greenwich. 39. M. Manufacturer. Enlisted Aug. 12, 1862. Mustered in Sept. 6, 1862. Transferred to Veteran Reserve Corps, Jan. 15, 1865. Died Mar. 20, 1894. Interred at St. Patrick's, East Greenwich, RI.

Bigelow, Charles H. Residence, Providence. 37. M. Laborer. Enlisted Aug. 22, 1862. Mustered in Sept. 6, 1862. Transferred to Veteran Reserve Corps Sept. 7, 1863. Interred at Grace Church Cemetery, Providence, RI.

Billington, Daniel R. Transferred from Co. G, Feb. 1, 1865. Mustered out June 9, 1865. Died 1927. Interred at Riverside Cemetery, South Kingstown, RI.

Briggs, Benjamin G. Residence, Newport. 35. M. Mason. Enlisted Aug. 22, 1862. Mustered in Sept. 6, 1862. Died of typhoid at Pleasant Valley, MD, Nov. 4, 1862.

Briggs, George W. Residence, Warwick. 25. S. Laborer. Enlisted Aug. 15, 1862. Mustered in Sept. 6, 1862. Deserted at Fort Wood, N. Y. Harbor, Jan. 23, 1865.

Briggs Lemuel A. Transferred from Co. G, Feb. 1, 1865. Mustered out June 9, 1865. Died July 18, 1901. Interred at Riverside Cemetery, South Kingstown, RI.

Briggs, Wanton S. Transferred from Co. G, Feb. 1, 1865. Mustered out June 9, 1865. Died April 1, 1879. Interred at Oak Grove Cemetery, Hopkinton, RI.

Boss, Joseph A. Transferred from Co. G, Feb. 1, 1865. Mustered out June 9, 1865. Died Mar. 9, 1934. Interred at Riverside Cemetery, South Kingstown, RI.

Brown, Samuel G. Residence, East Greenwich. 41. M. Carpenter. Enlisted Aug. 12, 1862. Mustered in Sept. 6, 1862. Promoted to corporal.

Browning, George T. Residence, Hopkinton. 24. S. Farmer. Enlisted Aug. 27, 1862. Mustered in Sept. 6, 1862. Wounded in action, shot in hand, at Bethesda Church, VA, June 3, 1864. Mustered out June 9, 1865.

Burke, John. Residence, Smithfield. 18. S. Laborer. Enlisted Aug. 12, 1862. Mustered in Sept. 6, 1862. Died of dysentery at Big Black River, MS, July 12, 1863.

Capron, Henry P. Residence, East Greenwich. 31. M. Teamster. Enlisted Aug. 15, 1862. Mustered in Sept. 6, 1862. Deserted in the face of the enemy at Fredericksburg, VA, Dec. 13, 1862. Gained from desertion Nov. 19, 1863, and sentenced by general court-martial to hard labor on public works for five years. Deserted at Camp Nelson, KY, Dec. 8, 1864.

Caswell, James D. Transferred from Co. G, Feb. 1, 1865. Sick in Mount Pleasant General Hospital, Washington, DC, where he was mustered out June 21, 1865. Died July 19, 1920. Interred at Riverside Cemetery, South Kingstown, RI.

Connor, Peter. Transferred from Co. G, Feb. 1, 1865. Mustered out June 9, 1865.

Conway, Patrick. Residence, East Greenwich. 24. S. Cigar maker. Enlisted Aug. 7, 1862. Mustered in Sept. 6, 1862. Wounded in action, shot in head, at Petersburg, VA, June 17, 1864. Mustered out June 9, 1865.

Cornell, Martin. Residence, Warwick. 31. M. Wheelwright. Enlisted Aug. 15, 1862. Mustered in Sept. 6, 1862. Wounded in action, shot in leg, at Fredericksburg, VA, Dec. 13, 1862. Mortally wounded in action at Spotsylvania Court House, VA, May 14, 1864. Died of wounds at Annapolis, MD, June 1, 1864. Interred at Large Maple Root Cemetery, Coventry, RI.

Covill, George W. Residence, Warwick. 19. S. Farmer. Enlisted and Mustered Jan. 26, 1864. Transferred to Co. D, June 9, 1865.

Dacy, John C. Residence, Providence. 33. M. Sailor. Enlisted Aug. 15, 1862. Mustered in Sept. 6, 1862. Deserted at Washington, DC, Sept. 13, 1862.

Dewhurst, Enoch. Residence, Warwick. 38. M. Shoemaker. Enlisted Aug. 15, 1862. Mustered in Sept. 6, 1862. Mustered out June 9, 1865. Died May 4, 1884. Interred at Swan Point Cemetery, Providence, RI.

Dexter, Henry R. Transferred from Co. G, Feb. 1, 1865. Mustered out June 9, 1865. Died Nov. 25, 1893. Interred at Grove Street Cemetery, Putnam, CT.

Donahue, Martin. Transferred from Co. G, Feb. 1, 1865. Mustered out June 9, 1865. Interred at Riverside Cemetery, South Kingstown, RI.

Donnelly, James. Residence, Providence. 21. S. Laborer. Enlisted July 22, 1862. Mustered in Sept. 6, 1862. Discharged for disability at Newport News, VA, Mar. 19, 1863.

Dowd, Oliver. Residence, East Greenwich. 21. S. Laborer. Enlisted Aug. 11, 1862. Mustered in Sept. 6, 1862. Wounded in action, shot in leg, at Fredericksburg, VA, Dec. 13, 1862. Discharged for disability at Newport News, Mar. 2, 1863. Died July 29, 1906. Interred at Knotty Oak Cemetery, Coventry, RI.

Effinger, Julius. Transferred from Co. G, Feb. 1, 1865. Discharged for disability at Alexandria, VA, June 24, 1865.

Fay, William. Residence, Westerly. 35. M. Laborer. Enlisted Aug. 15, 1862. Mustered in Sept. 6, 1862. Wounded in action at Spotsylvania Court House, VA, May 18, 1864. Transferred to Veteran Reserve Corps, Jan. 28, 1865.

Fitzgerald, Walter. Residence, Westerly. 29. M. Laborer. Enlisted Aug. 15, 1862. Mustered in Sept. 6, 1862. Wounded in action at Spotsylvania Court House, VA, May 18, 1864. Mustered out June 9, 1865. Died 1891. Interred at St. Michael's Cemetery, Stonington, CT.

Follett, Samuel O. Residence, East Greenwich. 19. S. Teamster. Enlisted Aug. 10, 1862. Mustered in Sept. 6, 1862. Mortally wounded in action at Spotsylvania Court House, VA, May 14, 1864. Died of wounds at Alexandria, VA, June 16, 1864. Interred at Alexandria National Cemetery. Grave 2166.

Foster, James A. Residence, Providence. Enlisted and Mustered Jan. 7, 1865. 19. S. Transferred to Co. D, June 9, 1865.

Fuller, Joseph A. Residence, North Providence. 30. M. Carder. Enlisted July 18, 1862. Mustered in Sept. 6, 1862. Mustered out June 9, 1865.

Gardiner, Henry. Residence, Westerly. 28. M. Teamster. Enlisted Aug. 14, 1862. Mustered in Sept. 6, 1862. Wagoner on extra duty in quartermaster's department from Dec. 1862, until Feb. 1863. Mustered out June 9, 1865.

Gavitt, Warren S. Residence, Westerly. 41. M. Machinist. Enlisted Aug. 15, 1862. Mustered in Sept. 6, 1862. Wounded in action, shot in knee, at Fredericksburg, VA, Dec. 13, 1862. Discharged for disability at Baltimore, MD, Feb. 27, 1863. Died 1875. Interred at Riverbend Cemetery, Westerly, RI.

Glavin, John. Residence, Westerly. 29. M. Laborer. Enlisted Aug. 15, 1862. Mustered in Sept. 6, 1862. Deserted at Pittsburg, PA, Mar. 29, 1863.

Gladding, Nathaniel W. Residence, Providence. 20. S. Machinist. Enlisted and Mustered Feb. 9, 1865. Transferred to Co. D, June 9, 1865.

Gorton, Richard, Jr. Residence, North Kingstown. 29. M. Manufacturer. Enlisted Aug. 15, 1862. Mustered in Sept. 6, 1862. Wounded in action at Fredericksburg, VA, Dec. 13, 1862. Sent to hospital and borne as absent sick until Feb. 1863. Killed in action at Spotsylvania Court House, VA, May 18, 1864. Interred at Fredericksburg National Cemetery. Grave 1165.

Gorton, Thomas. Residence, East Greenwich. 18. S. Laborer. Enlisted Aug. 12, 1862. Mustered in Sept. 6, 1862. Killed in action at Fredericksburg, VA, Dec. 13, 1862.

Gradwell, James. Residence, Warwick. 35. M. Printer. Enlisted Aug. 15, 1862. Mustered in Sept. 6, 1862. Wounded in action, shot in hand, at Spotsylvania Court House, VA, May 12, 1864. Mustered out June 9, 1865. Died Nov. 15, 1887. Interred at St. Phillip's Episcopal Cemetery, West Warwick, RI.

Greene, George D. Residence, Warwick. 19. S. Blacksmith. Enlisted Aug. 15, 1862. Mustered in Sept. 6, 1862. Captured at Fredericksburg, VA, Dec. 13, 1862. Paroled Dec. 17, 1862.

Mustered out June 9, 1865. Died 1933. Interred at Large Maple Root Cemetery, Coventry, RI.

Griffin, Joseph H. Transferred from Co. G, Feb. 1, 1865. Transferred to Co. K, Mar. 9, 1865.

Hall, William A. Residence, Richmond. 18. S. Laborer. Enlisted Aug. 7, 1862. Mustered in Sept. 6, 1862. On extra duty in hospital department Dec. 1862. Died of typhoid in hospital at Washington, DC, Feb. 10, 1863. Interred at Soldier's Home National Cemetery, Washington, DC. Grave 4374.

Harvey, James G. Residence, Providence. 26. S. Laborer. Enlisted July 8, 1863. Wounded in action, shot in leg, at Spotsylvania Court House, VA, May 12, 1864. Wounded in action at Spotsylvania Court House, VA, May 18, 1864. Transferred to Veteran Reserve Corps, Feb. 14, 1865. Died Jan. 30, 1921. Interred at Woodland Cemetery, Keene, NH.

Harvey, John Jr. Transferred from Co. G, Feb. 1, 1865. Mustered out June 9, 1865. Interred at Oak Dell Cemetery, South Kingstown, RI.

Harvey, William. Transferred from Co. G, Feb. 1, 1865. Mustered out June 9, 1865. Died May 17, 1905. Interred at Oak Dell Cemetery, South Kingstown, RI.

Hill, Baxter M. Residence, Providence. 19. S. Laborer. Enlisted and Mustered Jan. 25, 1865. Transferred to Co. D, June 9, 1865.

Hopkins, George L. Transferred from Co. G, Feb. 1, 1865. Mustered out June 9, 1865. Died June 19, 1904. Interred at North Burial Ground, Bristol, RI.

Hodson, James. Residence, Warwick. 38. M. Laborer. Enlisted Aug. 15, 1862. Mustered in Sept. 6, 1862. Killed in action at Cold Harbor, VA, June 6, 1864.

Holland, George A. Transferred from Co. G, Feb. 1, 1865. Mustered out June 9, 1865. Died 1907. Interred at Riverbend Cemetery, Westerly, RI.

Holland, George H. Transferred from Co. G, Feb. 1, 1865. Mustered out June 9, 1865. Died Nov. 10, 1918. Interred at Perryville Cemetery, South Kingstown, RI.

Holland, Reuben, Jr. Transferred from Co. G, Feb. 1, 1865. In division hospital until June 1865. Mustered out June 9, 1865. Interred at Oak Dell Cemetery, South Kingstown, RI.

Holloway, Elisha. Residence, Providence. 24. M. Clerk. Enlisted and Mustered Mar. 9, 1865. Transferred to Co. D, June 9, 1865.

Hopkins, Arnold. Residence, Warwick. 35. M. Laborer. Enlisted Aug. 15, 1862. Mustered in Sept. 6, 1862. Absent sick from Oct. 27, 1862, until Feb. 1863. Transferred to Veteran Reserve Corps, Jan. 5, 1864. Died Feb. 21, 1894. Interred at Manchester Cemetery, Coventry, RI.

Hull, James S. Residence, Westerly. 34. M. Operative. Enlisted Aug. 9, 1862. Mustered in Sept. 6, 1862. Wounded in action at Jackson, MS, July 13, 1863. Mustered out June 9, 1865. Died Mar. 26, 1906. Interred at Riverbend Cemetery, Westerly, RI.

Keating, Patrick. Residence, Providence. Mustered in Sept. 6, 1862. Deserted at Camp Chase, VA, Sept. 22, 1862.

Keaton, Daniel R. Transferred from Co. G, Feb. 1, 1865. Mustered out June 9, 1865. Interred at Togus National Cemetery.

Kimball, James. Residence, Newport. 30. M. Gunsmith. Enlisted Aug. 22, 1862. Mustered in Sept. 6, 1862. Mustered out June 9, 1865.

Knowles, Edward L. Residence, North Providence. 28. M. Silversmith. Enlisted Aug. 19, 1862. Mustered in Sept. 6, 1862. On duty in quartermaster's department from Nov. 1862, until Feb.

1863. Mustered out June 9, 1865. Died April 21, 1905. Interred at North Burial Ground, Providence, RI.

Leary, Jerry. Residence, Westerly. 22. S. Laborer. Enlisted Aug. 14, 1862. Mustered in Sept. 6, 1862. Killed in action at Fredericksburg, VA, Dec. 13, 1862.

Lee, Frank. Residence, Providence. 18. S. Teamster. Enlisted and Mustered Jan. 11, 1865. Transferred to Co. D, June 9, 1865.

Ledden, Daniel. Residence, North Providence. 18. S. Farmer. Enlisted Aug. 7, 1862. Mustered in Sept. 6, 1862. Killed in action at Fredericksburg, VA, Dec. 13, 1862.

Lovely, Alfred. Residence, Providence. 18. S. Spinner. Enlisted and Mustered Aug. 23, 1864. Mustered out June 9, 1865. Died Mar. 27, 1932. Interred at St. Mary Cemetery, Putnam, CT.

McFarland, Daniel. Residence, Pawtucket. 18. S. Laborer. Enlisted Aug. 13, 1862. Mustered in Sept. 6, 1862. Promoted to corporal June 20, 1863.

McKay, John, Jr. Residence, Warwick. 22. S. Clerk. Enlisted Aug. 30, 1862. Mustered in Sept. 6, 1862. Promoted to sergeant.

McKenna, Patrick. Residence, Westerly. 40. M. Farmer. Enlisted Aug. 16, 1862. Mustered in Sept. 6, 1862. Wounded in action, shot in left hand, at Fredericksburg, VA, Dec 13. 1862. Transferred to Veteran Reserve Corps. Died June 6, 1890. Interred at State Farm Cemetery, Cranston, RI.

Martin, John. Residence, Providence. 28. S. Blacksmith. Enlisted Nov. 7, 1863. Deserted at Annapolis, MD, April 26, 1864.

Marks, John. Residence, East Greenwich. 29. M. Shoemaker. Enlisted Aug. 15, 1862. Mustered in Sept. 6, 1862. Wounded in action, shot in hand, at Fredericksburg, VA, Dec. 13, 1862. Sent to hospital and borne as absent sick until Jan. 24, 1863, when he deserted at Hammond General Hospital, MD

Millard, James R. Residence, East Greenwich. 26. S. Peddler. Enlisted Aug. 15, 1862. Mustered in Sept. 6, 1862. On detached service in ambulance corps from Nov. 1862, until Feb. 1863. Absent sick Jan. 1865. Mustered out June 9, 1865.

Morrissey, John. Residence, Providence. 22. S. Laborer. Enlisted and Mustered Feb. 11, 1865. Transferred to Co. D, June 9, 1865.

Mowry, Benjamin. Residence, Warwick. 42. M. Laborer. Enlisted Sept. 1, 1862. Mustered in Sept. 6, 1862. Discharged for disability at Campbell General Hospital, Washington, DC, April 13, 1865. Died Nov. 13, 1896. Interred at Mowry-Knight-Mott Lott, Warwick Cemetery 37, Warwick, RI.

Mulholland, Hugh. Residence, Pawtucket. 29. M. Laborer. Enlisted Aug. 14, 1862. Mustered in Sept. 6, 1862. Deserted at Pittsburg, PA, Mar. 28, 1863.

Murphy, John T. Residence, Westerly. 36. M. Laborer. Enlisted Aug. 15, 1862. Mustered in Sept. 6, 1862. On extra duty in quartermaster's department from Dec. 1862, until Feb. 1863. Wounded in action at Spotsylvania Court House, VA, May 18, 1864. Mustered out June 9, 1865.

Nichols, Wanton A. Residence, East Greenwich. 21. S. Laborer. Enlisted Aug. 12, 1862. Mustered in Sept. 6, 1862. Discharged for disability at Portsmouth Grove, RI, Mar. 20, 1863.

Northrup, William H. Residence, North Providence. 25. M. Painter. Deserted from 11th R. I. Vols. Assigned to the 7th RI Vols. to make good time lost by desertion. Mustered out at Providence, RI, Feb. 13, 1865.

Olney, Zalmon A. Residence, Exeter. 40. M. Farmer. Enlisted Aug. 16, 1862. Mustered in Sept. 6, 1862. Killed in action at Fredericksburg, VA, Dec. 13, 1862.

Parker, Almond K. Residence, Richmond. 17. S. Mason. Enlisted Aug. 25, 1862. Mustered in Sept. 6, 1862. Discharged for disability at Camp Dennison, OH, Dec. 29, 1863. Died 1925. Interred at Wood River Cemetery, Richmond, RI.

Place, Arnold J. Residence, East Greenwich. 27. M. Laborer. Enlisted Aug. 12, 1862. Mustered in Sept. 6, 1862. Died of typhoid at Camp Nelson, KY, Feb. 26, 1864. Interred at Camp Nelson National Cemetery. Section D, Grave 1259.

Potter, Franklin H. Transferred from Co. G, Feb. 1, 1865. Mustered out June 9, 1865.

Potter, Jared J. Transferred from Co. G, Feb. 1, 1865. Mustered out June 9, 1865. Died April 4, 1890. Interred at White Brook Cemetery, Richmond, RI.

Rathbun, George. Residence, East Greenwich. 23. S. Farmer. Enlisted Aug. 11, 1862. Mustered in Sept. 6, 1862. Detached to Battery D, 1st R. I. Light Artillery from Jan. 15, 1863, until Dec. 10, 1864. Wounded in action at Cedar Creek, VA, Oct. 19, 1864. Mustered out June 9, 1865. Died Sept. 8, 1901. Interred at John Spencer Lot, East Greenwich Cemetery 9, East Greenwich, RI.

Rathbun, Nathan. Residence, East Greenwich. 30. S. Laborer. Enlisted Aug. 12, 1862. Mustered in Sept. 6, 1862. Mortally wounded in action at Jackson, MS, July 13, 1863. Died of wounds at U. S. Hospital, Covington, KY, Aug. 22, 1863. Interred at Camp Nelson National Cemetery. Section G, Grave 2012. Cenotaph at Small Maple Root Cemetery, Coventry, RI.

Rathbun, William. Residence, East Greenwich. 21. S. Laborer. Enlisted Aug. 12, 1862. Mustered in Sept. 6, 1862. Wounded in action, leg amputated, at Fredericksburg, VA, Dec. 13, 1862. Discharged for disability at Washington, DC, Mar. 6, 1863. Died July 17, 1906. Interred at William Potter Lot, West Greenwich Cemetery 30, West Greenwich, RI.

Rice, John E. Residence, Warwick. 23. S. Laborer. Enlisted Aug. 15, 1862. Mustered in Sept. 6, 1862. Wounded in action, shot in hand, at Jackson, MS, July 13, 1863. Wounded in action at Spotsylvania Court House, VA, May 12, 1864. Killed in action at

Spotsylvania Court House, VA, May 18, 1864. Interred at Fredericksburg National Cemetery. Grave 1159.

Rice, Michael. Residence, Smithfield. 22. S. Laborer. Enlisted Aug. 18, 1862. Mustered in Sept. 6, 1862. Transferred to Veteran Reserve Corps May 1, 1864.

Scott, Bradford W. Residence, Coventry. 28. M. Teamster. Enlisted Aug. 13, 1862. Mustered in Sept. 6, 1862. Mustered out June 9, 1865. Died May 12, 1898. Interred at Joseph Scott Cemetery, Coventry Cemetery 139, Coventry, RI.

Scott, Walter R. Residence, Sterling, CT. 34. M. Farmer. Enlisted Aug. 15, 1862. Mustered in Sept. 6, 1862. Died of dysentery at U. S. General Hospital, Covington, KY, Aug. 19, 1863. Interred at Riverside Cemetery, Sterling, CT.

Shippee, Albert G. Residence, East Greenwich. 42. M. Farmer. Enlisted Aug. 12, 1862. Mustered in Sept. 6, 1862. Absent sick at Pleasant Valley, MD, from Oct. 27, 1862, until Feb. 1863. Discharged for disability at Camp Dennison, OH, Dec. 16, 1863. Died Feb. 23, 1904. Interred at Shippee-Arnold Lot, East Greenwich Cemetery 5, East Greenwich, RI.

Smith, James. Residence, Providence. S. 18. Laborer. Enlisted and Mustered Feb. 10, 1865. Transferred to Co. D, June 9, 1865.

Stillwell, Allen G. Residence, Providence. 28. S. Clerk. Enlisted July 10, 1862. Mustered in Sept. 6, 1862. Discharged for disability at Washington, DC, Feb. 14, 1863. Died Feb. 25, 1909. Interred at Riverbend Cemetery, Westerly, RI.

Stone, Albert. Transferred from Co. G, Feb. 1, 1865. Mustered out June 9, 1865. Died July 23, 1899. Interred at Elm Grove Cemetery, North Kingstown, RI.

Sweet, Albert. Residence, Providence. 44. M. Hair dresser. Enlisted July 21, 1862. Mustered in Sept. 6, 1862. Transferred to Veteran Reserve Corps, May 1, 1864. Died 1903. Interred at Cook Cemetery, Woonsocket, RI.

Sweet, Charles E. Residence, Warwick. 25. M. Farmer. Enlisted Aug. 15, 1862. Mustered in Sept. 6, 1862. Mustered out June 9, 1865. Died Dec. 31, 1914. Interred at Glenwood Cemetery, East Greenwich, RI.

Sweet, James W. Residence, Warwick. 20. S. Farmer. Enlisted Aug. 15, 1862. Mustered in Sept. 6, 1862. Discharged for disability at Falmouth, VA, Dec. 9, 1863. Died April 21, 1896. Interred at Glenwood Cemetery, East Greenwich, RI.

Sweet, John B. Residence, Bristol. 24. M. Merchant. Enlisted Aug. 21, 1862. Mustered in Sept. 6, 1862. Wounded in action, shot in shoulder, at Fredericksburg, VA, Dec. 13, 1862. Wounded in action, shot in leg, at Spotsylvania Court House, VA, May 12, 1864. Sent to hospital and borne as absent sick until July 7, 1864, when he returned to duty. Mustered out June 9, 1865. Interred at Moshassuck Cemetery, Central Falls, RI.

Sweet, John C. Residence, Warwick. 27. M. Laborer. Enlisted Aug. 15, 1862. Mustered in Sept. 6, 1862. Discharged for disability at Pleasant Valley, MD, Oct. 25, 1862.

Sweet, Gardiner C. Residence, Pawtucket. 43. M. Laborer. Enlisted Aug. 16, 1862. Mustered in Sept. 6, 1862. Wounded in action at Fredericksburg, VA, Dec. 13, 1862. Discharged for disability at Baltimore, MD, Oct. 12, 1863.

Taylor, Isaac Y. Residence, North Kingstown. 23. S. Farmer. Enlisted Aug. 16, 1862. Mustered in Sept. 6, 1862. Promoted to corporal June 20, 1863.

Taylor, James J. Residence, Warwick. 28. M. Laborer. Enlisted Aug. 15, 1862. Mustered in Sept. 6, 1862. Wounded in action at Cold Harbor, VA, June 8, 1864. Wounded in action, at Petersburg, VA, June 18, 1864. Mustered out June 9, 1865. Died 1906. Interred at Glenwood Cemetery, East Greenwich, RI.

Taylor, Stephen P. Residence, Warwick. 20. S. Farmer. Enlisted Aug. 15, 1862. Mustered in Sept. 6, 1862. Died of typhoid at

Annapolis, MD, April 13, 1864. Interred at First Cemetery, East Greenwich, RI.

Tefft, Samuel S. Transferred from Co. G, Feb. 1, 1865. Absent sick in hospital at Washington, DC, until mustered out May 31, 1865. Died June 7, 1899. Interred at Riverside Cemetery, South Kingstown, RI.

Wells, Benjamin E. Transferred from Co. G. Mustered out June 9, 1865. Died Nov. 5, 1917. Interred at Oak Grove Cemetery, Fall River, MA.

Wilson, William R. Residence, Warwick. 30. M. Peddler. Enlisted Aug. 15, 1862. Mustered in Sept. 6, 1862. Discharged for disability at Newport News, VA, Mar. 19, 1863. Died Nov. 30, 1868. Interred at First Cemetery, East Greenwich, RI.

Young, George W. Residence, Providence. 20. S. Laborer. Enlisted Aug. 8, 1862. Mustered in Sept. 6, 1862. Discharged for disability at Portsmouth Grove, RI, Mar. 16, 1864.

COMPANY I

Captains

Carr, Thomas H. Residence, Newport. M. 41. Blacksmith. Commissioned Sept. 4, 1862. Mustered in Sept. 6, 1862. Resigned Jan. 10, 1863. Died March 18, 1897. Interred at Island Cemetery, Newport, RI.

Jenks, Ethan Amos. Promoted from second lieutenant Co. H, Mar. 3, 1863. Brevet major for gallantry at Spotsylvania Court House, VA, May 18, 1864. Wounded in action, shot in leg, at Petersburg, VA, June 19, 1864. Commissioned major June 29, 1864.

Kenyon, David R. Promoted from first lieutenant Co. A, Jan. 7, 1863. Resigned Mar. 2, 1863. Died July 14, 1897. Interred at Wood River Cemetery, Richmond, RI.

First Lieutenants

Brownell, Thomas S. Residence, Newport. M. Tinsmith. Commissioned Sept. 4, 1862. Mustered in Sept. 6, 1862. Resigned Jan. 1, 1863. Died 1901. Interred at South Burial Ground, Warren, RI.

Hunt, Edwin L. Promoted from second lieutenant Co. I, Mar. 1, 1863. Promoted to captain Co. G, May 3, 1864.

McIlroy, Samuel. Promoted from second lieutenant Co. I, July 25, 1864. Mortally wounded in action Sept. 30, 1864 at Poplar Spring Church, VA. Died of wounds Oct. 25, 1864 at City Point, VA. Interred at Mineral Spring Cemetery, Pawtucket, RI.

Merrill, James F. Transferred from Co. D, Feb. 1, 1865. Mustered out June 9, 1865.

Phelps, James T. Promoted from second lieutenant Co. I, Nov. 25, 1864, but not mustered until Jan. 1, 1865. Transferred to Co. H, Feb. 1, 1865.

Second Lieutenants

Cheever, John T. H. Residence, Providence. 29. S. Commissioned Mar. 2, 1863. Never reported to company. On special duty on General Silas Casey's staff at Washington, DC. Resigned Dec. 26, 1863. Died of illness contracted in the service June 6, 1865 at Wrentham, MA. Interred at Gerould Cemetery, Wrentham, MA.

Hunt, Edwin L. Promoted from private Co. I, Oct. 24, 1862. Promoted to first lieutenant Co. I, Mar. 1, 1863.

McIlroy, Samuel. Promoted from sergeant Co. I, May 12, 1864. Wounded in action, shot in thigh, at Bethesda Church, VA, June 3, 1864. Promoted to first lieutenant Co. I, July 25, 1864.

Phelps, James T. Promoted from sergeant, Co. I, July 25, 1864. Promoted to first lieutenant Co. I, Nov. 25, 1864.

Smith, Albert L. Promoted from sergeant Co. F, Mar. 1, 1863. Mustered in Mar. 17, 1863. Promoted to first lieutenant Co. D, April 3, 1863.

First Sergeants

Deane, Arthur W. Promoted from private Feb. 1, 1863. Mustered out June 9, 1865. Died Sept. 13, 1909. Interred at Greenwood Cemetery, Coventry, RI.

McKee, J. Frank. Residence, Newport. 25. M. Painter. Enlisted, Aug. 19, 1862. Mustered in Sept. 6, 1862. Discharged for disability at Convalescent Camp, VA, Jan. 31, 1863.

Sergeants

Danforth, George A. Transferred from Co. D, Feb. 1, 1865. Mustered out June 9, 1865. Died Jan. 6, 1869. Interred at North Burial Ground, Providence, RI.

Dingley, Fuller. Residence, Newport. 30. M. Carpenter. Enlisted Aug. 13, 1862. Mustered in Sept. 6, 1862. Wounded in action, shot in arm, at Fredericksburg, VA, Dec. 13, 1862. Borne as absent sick until Feb. 1863. Promoted to second lieutenant Co. D, May 20, 1863.

Lincoln, Henry. Promoted from private Sept. 16, 1862. Promoted to second lieutenant Co. C, Jan. 7, 1863.

McIlroy, Samuel. Promoted from corporal. Wounded in action, in head, at Fredericksburg, VA, Dec. 13, 1862. Promoted to second lieutenant Co. I, May 12, 1864.

Mayo, Charles H. V. Residence, Bristol. 19. S. Printer. Enlisted Aug. 15, 1862. Mustered in Sept. 6, 1862. Wounded in action, shot in right heel, at Fredericksburg, VA, Dec. 13, 1862. Transferred to Veteran Reserve Corps, Nov. 20, 1863. Died March 24, 1905. Interred at North Burial Ground, Providence, RI.

Miller, Benjamin F. Promoted from private. Wounded in action, in left eye, at Fredericksburg, VA, Dec. 13, 1862. Wounded in action at the Wilderness, VA, May 5, 1864. Mortally wounded in action, shot in hip, June 3, 1864 at Bethesda Church, VA. Sent to Harewood Hospital, Washington, DC. Died of wounds at Providence, RI, Nov. 11, 1866. Interred at North Burial Ground, Providence, RI.

Morse, Ephraim C. Residence, Newport. 40. M. Carpenter. Enlisted Aug. 13, 1862. Mustered in Sept. 6, 1862. Wounded in action, in head, at Fredericksburg, VA, Dec. 13, 1862. Promoted to second lieutenant Co. G, April 3, 1863.

Phelps, James T. Promoted from corporal. Wounded in action, shot in finger at Bethesda Church, VA, June 3, 1864. Wounded in action, shot in hand, at Petersburg, VA, June 16, 1864. Promoted to second lieutenant Co. I, July. 25, 1864.

Scannell, Dennis J. Promoted from private. Sick at Portsmouth Grove, RI, from Jan. until June 1865. Mustered out June 15, 1865. Died Feb. 10, 1876. Interred at St. Johns Cemetery, Worcester, MA.

Shumway, Amos D. Transferred from Co. D, Feb. 1, 1865. Mustered out June 9, 1865. Interred at Mt. Zion Cemetery, Webster, MA.

Simpson, Samuel F. Promoted from corporal. Served as regimental color bearer. Killed in action at the North Anna River, VA, May 25, 1864.

Sprague, John H. D. Transferred from Co. D, Feb. 1, 1865. Mustered out June 9, 1865. Died Sept. 28, 1900. Interred at Hilldale, NJ.

Stanhope, John R. Jr. Residence, Newport. 39. M. Grocer. Enlisted Aug. 13, 1862. Mustered in Sept 6, 1862. Transferred to non-commissioned staff as quartermaster sergeant, Sept. 7, 1862.

Corporals

Anthony, Edward, Jr. Residence, Bristol. 22. S. Farmer. Enlisted Aug. 15, 1862. Mustered in Sept. 6, 1862. Discharged for disability at Frederick, MD, Dec. 29, 1862. Died March 11, 1914. Interred at Juniper Hill Cemetery, Bristol, RI.

Arnold, Israel B. Promoted from private. Wounded in action, right toe amputated, at Fredericksburg, VA, Dec. 13, 1862. Sent to Portsmouth Grove, RI until Feb. 1865. Assigned by order to duty as hospital steward in Post Medical Department, Louisville, KY, Feb. 6, 1865, and relieved from duty June 15, 1865. Mustered out at Louisville, KY, June 16, 1865. Died Jan. 27, 1871. Interred at Pawtuxet Cemetery, Warwick, RI.

Clark, Erasmus D. Residence, Newport. 20. S. Cabinet maker. Enlisted Aug. 11, 1862. Mustered in Sept. 6, 1862. Mustered out July 25, 1865. Died Jan. 19, 1919. Interred at Common Burying Ground, Newport, RI.

Carter, Gideon W. Promoted from private. Mustered out June 9, 1865. Died Nov. 4, 1918. Interred at Westfield Cemetery, Danielson, CT.

Darling, Esek R. Transferred from Co. D, Feb. 1, 1865. Mustered out June 9, 1865. Died Nov. 9, 1930. Interred at Pascoag Cemetery, Burrillville, RI.

Hanning, Robert. Promoted from private. Wounded in action, shot in chest at Bethesda Church, VA, June 3, 1864. Mustered out June 9, 1865. Served in Color Guard.

Jones, Peleg G. Promoted from private. Mustered out June 9, 1865. Died 1902. Interred at North Burial Ground, Bristol, RI.

McIlroy, Samuel. Residence, Pawtucket. 36. M. Engraver. Enlisted June 10, 1862. Mustered in Sept. 6, 1862. Promoted to sergeant.

Mulligan, Henry A. Residence, Pawtucket. 42. M. Carpenter. Enlisted Aug. 12, 1862. Mustered in Sept. 6, 1862. Discharged for disability at Falmouth, VA, Jan. 17, 1863.

Nicholas, James A. Residence, Cranston. 36. M. Spinner. Enlisted Aug. 13, 1862. Mustered in Sept. 6, 1862. In division hospital Mar. 1865, and so borne until June 1865. Mustered out June 9, 1865. Died Dec. 19, 1904. Interred at Pocasset Cemetery, Cranston, RI.

Phelps, James T. Residence, Bristol. 21. S. Jeweler. Enlisted Aug. 15, 1862. Mustered in Sept. 6, 1862. Promoted to sergeant.

Simpson, Samuel F. Residence, Newport. 42. M. Farmer. Enlisted Aug. 5, 1862. Mustered in Sept. 6, 1862. Promoted to sergeant.

Spooner, Charles D. Promoted from private. Mustered out June 9, 1865. Served in Color Guard. Died Sept. 2, 1899. Interred at State Farm Cemetery, Cranston, RI.

Walker, William H. Promote from private. Discharged for disability Sept. 25, 1863.

Whitman, Olney A. Residence, Newport. 33. M. Spinner. Enlisted Aug. 12, 1862. Mustered in Sept. 6, 1862. Died of typhoid at Baltimore, MD, Mar. 30, 1863. Interred at Peckham West Cemetery, New Bedford, MA.

Musicians

Hopkins, William P. Transferred from Co. D, Feb. 1, 1865. Mustered out June 9, 1865. Died Oct. 9, 1920. Interred at Knotty Oak Cemetery, Coventry, RI.

Morris, Joseph N. Residence, Bristol. 20. S. Cooper. Enlisted Aug. 1, 1862. Mustered in Sept. 6, 1862. Mustered out June 9, 1865. Died Oct. 23, 1895. Interred at North Burial Ground, Bristol, RI.

Sherman, Carder H. Transferred from Co. G, Feb. 1, 1865. Mustered out June 9, 1865. Died May 19, 1919. Interred at Oak Dell Cemetery, South Kingstown, RI.

Privates

Adams, Thomas J. Residence, Newport. 24. M. Spinner. Enlisted Aug. 13, 1862. Mustered in Sept. 6, 1862. Wounded in action, shot in right arm, at Fredericksburg, VA, Dec. 13, 1862, and sent to hospital. Discharged for disability at Portsmouth Grove, RI, Sept. 17, 1863. Died April 30, 1928. Interred at Hope Cemetery, Worcester, MA.

Allen, James. Residence, Providence. 21. S. Laborer. Enlisted Aug. 19, 1862. Mustered in Sept. 6, 1862. Mustered out June 9, 1865. Died Sept. 24, 1897. Interred at Swan Point Cemetery, Providence, RI.

Arnold, Israel B. Residence, Warwick. 32. S. Farmer. Enlisted Aug. 19, 1862. Mustered in Sept. 6, 1862. Promoted to corporal.

Barker, Alexander. Residence, Newport. 21. S. Painter. Enlisted Aug. 11, 1862. Mustered in Sept. 6, 1862. Wounded in action, in right side, arm, and head, at Fredericksburg, VA, Dec. 13, 1862, and sent to hospital. Discharged for disability from Emory Hospital, Washington DC, Feb. 3, 1863. Died of disease contacted in the service August 5, 1866 at Newport, RI. Interred at Common Burying Ground, Newport, RI.

Beard, Edward. Residence, East Providence. 32. M. Laborer. Enlisted Aug. 12, 1862. Mustered in Sept. 6, 1862. Transferred to the Veteran Reserve Corps, Feb. 15, 1864. Died 1920. Interred at Lakeside-Carpenter Cemetery, East Providence, RI.

Beckford, George C. Transferred from Co. D, Feb. 1, 1865. Mustered out June 9, 1865. Interred at Oak Grove Cemetery, Pawtucket, RI.

Beaumont, Ralph. Transferred from Co. D, Feb. 1, 1865. Mustered out June 9, 1865. Died 1916. Interred at North Cemetery, Oxford, MA.

Bliss, Lewis S. Residence, Bristol. 32. M. Laborer. Enlisted Aug. 16, 1862. Mustered in Sept. 6, 1862. Wounded in action at Petersburg, VA, June 15, 1864. In ambulance corps Jan. 1865, and so borne until April 1865. Mustered out June 9, 1865. Died Dec. 21, 1888. Interred at North Burial Ground, Bristol, RI.

Bullock, Eben F. Residence, Smithfield. 31. M. Weaver. Enlisted Aug. 15, 1862. Mustered in Sept. 6, 1862. Discharged for disability Mar. 17, 1863. Interred at North Burial Ground, Bristol, RI.

Bullock, Norman. Residence, Providence. 27. M. Butcher. Enlisted Aug. 12, 1862. Mustered in Sept. 6, 1862. Teamster in quartermaster's department Nov. 1862. Discharged for disability Mar. 19, 1863. Died Dec. 4, 1915. Interred at Evergreen Cemetery, Vinton, IA.

Branigan, John B. Transferred from Co. D, Feb. 1, 1865. Mustered out at Providence, June 20, 1865. Died Aug. 13, 1913. Interred at North Burial Ground, Providence, RI.

Brown, Marcus M. Transferred from Co. D, Feb. 1, 1865. Mustered out June 9, 1865. Died May 14, 1934. Interred at North Burial Ground, Bristol, RI.

Burgess, John H. Residence, Newport. 36. M. Farmer. Enlisted Aug. 15, 1862. Mustered in Sept. 6, 1862. Mustered out June 9, 1865. Interred at Island Cemetery, Newport, RI.

Card, James T. Residence, Bristol. 22. S. Cooper. Enlisted Aug. 19, 1862. Mustered in Sept. 6, 1862. Mustered out June 9, 1865. Died 1918. Interred at North Burial Ground, Bristol, RI.

Carragan, Martin W. Transferred from Co. D, Feb. 1, 1865. Mustered out June 9, 1865.

Carroll, John. Residence, Newport. 23. S. Cabinet Maker. Enlisted Aug. 15, 1862. Mustered in Sept. 6, 1862. Promoted to sergeant and transferred to Co. C, Mar. 10, 1863.

Carter, Gideon W. Residence, Providence. 31. M. Weaver. Enlisted Aug. 16, 1862. Mustered in Sept. 6, 1862. Promoted to corporal.

Chafee, Thomas D. Residence, Bristol. 34. M. Tailor. Enlisted Aug. 19, 1862. Mustered in Sept. 6, 1862. In commissary department April 1865. Mustered out June 9, 1865. Interred at North Burial Ground, Bristol, RI.

Chafee, Stephen B. Residence, Bristol. 23. S. Farmer. Enlisted Aug. 15, 1862. Mustered in Sept. 6, 1862. Mustered out June 9, 1865. Died Sept. 15, 1910. Interred at Indian Hill Cemetery, Middletown, CT.

Claflin, William H. Residence, Newport. 37. M. Carrier. Enlisted Aug. 15, 1862. Mustered in Sept 6, 1862. Mustered out June 9, 1865.

Colley, Thomas. Residence, Barrington. 34. M. Oysterman. Enlisted Aug. 19, 1862. Mustered in Sept. 6, 1862. Deserted in the face of the enemy at Fredericksburg, VA, Dec. 13, 1862. Died Mar. 6, 1890. Interred at Prince's Hill Cemetery, Barrington, RI.

Collins, William. Residence, Bristol. 27. M. Laborer. Enlisted Aug. 19, 1862. Mustered in Sept. 6, 1862. Wounded in action, shot in left shoulder, at Fredericksburg, VA, Dec. 13, 1862. Detached temporarily to Battery D, 1st RI Light Artillery, Jan. 15, 1863. Died of dysentery in Asylum Hospital, Knoxville, TN, May 1, 1864. Interred at Knoxville National Cemetery. Section D, Grave 758.

Cook, Job H. Residence, Portsmouth. 19. S. Farmer. Enlisted Aug. 15, 1862. Mustered in Sept. 6, 1862. Deserted at York, PA, Mar. 27, 1863. Died Nov. 30, 1869. Interred at Greenwood Cemetery, Brooklyn, NY.

Dakin, William H. Residence, Newport. 32. M. Shoemaker. Enlisted Aug. 16, 1862. Mustered in Sept. 6, 1862. Transferred to Veteran Reserve Corps, May 31, 1864.

Dawley, Varnum H. Transferred from Co. D, Feb. 1, 1865. Mustered out June 9, 1862. Died 1910. Interred at Pocasset Cemetery, Cranston, RI.

Deane, Arthur W. Residence, Coventry. 23. M. Baker. Enlisted Aug. 19, 1862. Mustered in Sept. 6, 1862. Promoted to first sergeant Feb. 1, 1863.

Denicoe, Frank, Jr. Transferred from Co. D, Feb. 1, 1865. Transferred to the Veteran Reserve Corps Mar. 22, 1865.

Denicoe, Joseph. Transferred from Co. D, Feb. 1, 1865. Mustered out June 9, 1865

Eddy, Daniel D. Residence, Newport. 20. S. Machinist. Enlisted Aug. 12, 1862. Mustered in Sept. 6, 1862. Temporarily detached to Battery E, 4th U. S. Artillery, Oct. 11, 1862. Discharged for disability at Fort Washington, MD, Jan. 16, 1865. Interred at Clayville Cemetery, Foster, RI.

Fagan, Patrick. Transferred from Co. D, Feb. 1, 1865. Mustered out June 9, 1865.

Fitts, Thomas. Residence, Bristol. 26. S. Shoemaker. Enlisted Aug. 16, 1862. Mustered in Sept. 6, 1862. Transferred to Veteran Reserve Corps Jan. 5, 1864.

Folsom, William H. Residence, Newport. 21. S. Printer. Enlisted Aug. 19, 1862. Mustered in Sept. 6, 1862. Mustered out June 9, 1865. Died 1885. Interred at Worcester, MA.

Franklin, Josephus. Residence, Bristol. 21. S. Hostler. Enlisted Aug. 16, 1862. Mustered in Sept. 6, 1862. Died of typhoid at Falmouth, VA, Nov. 29, 1862. Interred at North Burial Ground, Bristol, RI.

Gardner, Francis W. Residence, Smithfield. 39. M. Farmer. Enlisted Aug. 19, 1862. Mustered in Sept. 6, 1862. Died of Yazoo Fever at Camp Nelson, KY, Aug. 28, 1863.

Garey, John W. Residence, Newport. 21. S. Oysterman. Enlisted Aug. 13, 1862. Mustered in Sept. 6, 1862. Wounded in action, shot in thigh and abdomen, at Fredericksburg, VA, Dec. 13, 1862. Discharged for disability at Armory Square Hospital, Washington, DC, Feb. 18, 1863. Died Jan. 7, 1883. Interred at Island Cemetery, Newport, RI.

Gibney, Charles P. Residence, Warwick. 26. S. Jeweler. Enlisted Aug. 19, 1862. Mustered in Sept. 6, 1862. Mustered out June 9, 1865. Interred at Oakland Cemetery, Cranston, RI.

Gladding, James H. Residence, Bristol. 18. S. Laborer. Enlisted Aug. 16, 1862. Mustered in Sept. 6, 1862. Wounded in action at Fredericksburg, VA, Dec. 13, 1862. Wounded in action at Spotsylvania Court House, VA, May 12, 1864. Mortally wounded in action, shot in arm, at Bethesda Church, VA, June 3, 1864. Died of wounds at Mount Pleasant Hospital, Washington, DC, July 3, 1864. Interred at North Burial Ground, Bristol, RI.

Greene, Charles T. Residence, South Kingstown. 19. S. Student. Enlisted and Mustered in Sept. 6, 1862. Wounded in action, shot in eye and leg, at Fredericksburg, VA, Dec. 13, 1862. Discharged for disability Mar. 27, 1863.

Green, Benjamin F. Residence, Cranston. 18. S. Farmer. Enlisted Aug. 16, 1862. Mustered in Sept 6, 1862. Discharged for disability at Frederick, MD, Dec. 9, 1862.

Gomes, Frank P. Residence, Newport. 25. S. Carpenter. Enlisted Aug. 11, 1862. Mustered in Sept. 6, 1862. Sick at Pleasant Valley, MD, from Oct. 27, 1862, until Dec. 1862. Transferred to Veteran Reserve Corps, Mar. 7, 1864. Died June 28, 1924. Interred at Common Burying Ground, Newport, RI.

Hanning, Robert. Residence, Newport. 26. S. Coachman. Enlisted Aug. 12, 1862. Mustered in Sept. 6, 1862. Promoted to corporal.

Harrington, John Jr. Residence, Cranston. 27. M. Laborer. Enlisted Aug. 13, 1862. Mustered in Sept. 6, 1862. Deserted at Falmouth, VA, Jan. 24, 1863.

Hathaway, Alvin P. Residence, Newport. 40. M. Farmer. Enlisted Aug. 13, 1862. Mustered in Sept. 6, 1862. Mortally wounded in action at Petersburg, VA, June 24, 1864. Died of wounds at Mount Pleasant Hospital, Washington, DC, June 28, 1864. Interred at Arlington National Cemetery. Grave 5555.

Hayes, Samuel A. Transferred from Co. D, Feb. 1, 1865. Mustered out June 9, 1865.

Hill, Charles E. Residence, Newport. 24. S. Gunsmith. Enlisted Aug. 18, 1862. Mustered in Sept.6, 1862. Mustered out June 9, 1865. Died August 22, 1897. Interred at South Vershire Cemetery, Vershire, VT.

Hoard, James, Jr. Residence, Bristol. 19. S. Hostler. Enlisted Aug. 16, 1862. Mustered in Sept. 6, 1862. Wounded in action, right arm amputated, at Bethesda Church, VA, June 3, 1864. Discharged for disability at General Hospital, New York, NY, April 3, 1865. Died July 29, 1907. Interred at North Burial Ground, Bristol, RI.

Horton, Theodore. Residence, Providence. Enlisted Aug. 13, 1862. Mustered in Sept. 6, 1862. Wounded in action at Fredericksburg, VA, Dec. 13, 1862. Deserted at Portsmouth Grove, RI, Aug. 14, 1863.

Humes, Charles. Transferred from Co. D, Feb. 1, 1865. Mustered out June 9, 1865. Died Nov. 20, 1928. Interred at Plains Cemetery, Canterbury, CT.

Humes, Emory. Transferred from Co. D, Feb. 1, 1865. Mustered out June 9, 1865. Died Dec. 31, 1921. Interred at Pine Grove Cemetery, Waterboro, ME.

Hunt, Edwin L. Residence, Newport. 27. S. Clerk. Enlisted Aug. 18, 1862. Mustered in Sept. 6, 1862. Promoted to second lieutenant Co. I, Oct. 24, 1862.

Johnson, William H. Residence, Newport. 27. S. Photographer. Enlisted Aug. 15, 1862. Mustered in Sept. 6, 1862. Transferred to Co. D, Jan. 29, 1863.

Jones, David G. Residence, Providence. 23. S. Shoemaker. Enlisted Aug. 13, 1862. Mustered in Sept. 6, 1862. Wounded in action, shot in throat, at Fredericksburg, VA, Dec. 13, 1862. Discharged for disability Feb. 4, 1863.

Jones, Peleg G. Residence, Bristol. 31. M. Carpenter. Enlisted Aug. 15, 1862. Mustered in Sept. 6, 1862. Promoted to corporal.

Josyln, Benjamin. Transferred from Co. D, Feb. 1, 1865. Mustered out June 9, 1865.

Irons, Charles A. S. Transferred from Co. D, Feb. 1, 1865. Mustered out June 9, 1865. Died Oct. 31, 1889. Interred at Pascoag Cemetery, Burrillville, RI.

Kelley, Owen. Residence, Providence. 25. S. Jeweler. Enlisted Jan. 17, 1865. Transferred to Co. G, June 9, 1865.

Kilroy, John. Residence, Newport. 42. M. Laborer. Enlisted Aug. 12, 1862. Mustered in Sept. 6, 1862. Killed in action at Petersburg, VA, June 30, 1864.

King, George H. Residence, Bristol. 26. M. Laborer. Enlisted Aug. 15, 1862. Mustered in Sept. 6, 1862. Wounded in action at Fredericksburg, VA, Dec. 13, 1862. Discharged for disability at Providence, RI, Mar. 16, 1863. Died 1917. Interred at North Burial Ground, Bristol, RI.

Knight, Elisha C. Residence, Coventry. 26. M. Laborer. Enlisted Aug. 13, 1862. Mustered in Sept. 6, 1862. Wounded in action, shot in hand, at Bethesda Church, VA, June 3, 1864. Wounded in action, at Petersburg, VA, June 16, 1864. Wounded in action, shot in hand, at Petersburg, VA, June 22, 1864. Mustered out June 9, 1865. Died May 29, 1917. Interred at Pine Grove Cemetery, Coventry, RI.

Langworthy, George A. Residence, Hopkinton. 27. M. Grocer. Enlisted Aug. 12, 1862. Mustered in Sept. 6, 1862. Wounded in action, struck by shell, at Petersburg, VA, April 2, 1865. Mustered out June 9, 1865. Died Mar. 26, 1901. Interred at Oak Grove Cemetery, Hopkinton, RI.

Lee, Cornelius. Transferred from Co. D, Feb. 1, 1865. Mustered out June 9, 1865. Interred at Old Calvary Cemetery, Boston, MA.

Lincoln, Henry. Residence, Cranston. 25, M. Soldier. Enlisted Aug. 12, 1862. Mustered in Sept. 6, 1862. Promoted to sergeant Sept. 16, 1862.

Locklin, Thomas, Jr. Transferred from Co. D, Feb. 1, 1865. Transferred to Co. G, June 9, 1865.

Lyons, Luke. Residence, Providence. 21. S. Blacksmith. Enlisted Aug. 18, 1862. Mustered in Sept. 6, 1862. Wounded in action, shot in chest at Jackson, MS, July 13, 1863. Mustered out June 9, 1865. Died Sept. 13, 1888. Interred at St. Francis Cemetery, Pawtucket, RI.

McCann, Daniel A. Residence, Newport. 22. S. Printer. Enlisted Aug. 19, 1862. Mustered in Sept. 6, 1862. On detached service at Louisville, KY, Jan. 1865. Mustered out at Philadelphia, PA, May 19, 1865. Died June 19, 1908. Interred at North Burial Ground, Bristol, RI.

McGarvey, John. Residence, Providence. 42. M. Oysterman. Enlisted Aug. 18, 1862. Mustered in Sept. 6, 1862. Absent sick at City Point, Mar.1865. Mustered out June 9, 1865. Died Oct. 3, 1873. Interred at Newman Cemetery, East Providence, RI.

McNaulty, Hugh. Transferred from Co. D, Feb. 1, 1865. Mustered out June 9, 1865.

Manchester, Alexander H. 19. S. Laborer. Residence, Bristol. Enlisted Aug. 18, 1862. Mustered in Sept. 6, 1862. Mortally wounded in action at Bethesda Church, VA, June 3, 1864. Died of wounds at Washington, D C, June 15, 1864. Interred at North Burial Ground, Bristol, RI.

Manchester, Isaac B. Residence, Bristol. 21. S. Laborer. Enlisted Aug. 18, 1862. Mustered in Sept. 6, 1862. Died of typhoid at Armory Square Hospital, Washington, DC, Dec. 1, 1862. Interred at North Burial Ground, Bristol, RI.

Meigs, John R. Residence, Bristol. 24. M. Farmer. Enlisted Aug. 15, 1862. Mustered in Sept. 6, 1862. Absent sick at Alexandria, VA, Jan. 1865. Discharged for disability at Fairfax Seminary, VA, Mar. 20, 1865. Died July 1, 1890. Interred at North Burial Ground, Bristol, RI.

Miller, Benjamin F. Residence, North Providence. 26. M. Carpenter. Enlisted Aug. 1, 1862. Mustered in Sept. 6, 1862. Promoted to sergeant.

Mott, Caleb, Jr. Residence, Warwick. 19. S. Silversmith. Enlisted Aug. 12, 1862. Mustered in Sept. 6, 1862. Wounded in action, shot in thigh, at Fredericksburg, VA, Dec. 13, 1862. Sent to hospital and borne as absent sick until Feb. 1863. Transferred to the Veteran Reserve Corps, Nov. 20, 1863. Died May 22, 1871. Interred at Mowry-Knight-Mott Lot, Warwick Cemetery 37, Warwick, RI.

Moses, Ashael O. Residence, Cranston. 39. M. Mason. Enlisted Aug. 21, 1862. Mustered in Sept. 6, 1862. Deserted at Washington, DC, Sept. 13, 1862.

Munroe, Francis. Residence, Bristol. 26. S. Farmer. Enlisted Aug. 15, 1862. Mustered in Sept. 6, 1862. Transferred to the Veteran Reserve Corps, Jan. 15, 1864.

Nichols, Daniel. Transferred from Co. D, Feb. 1, 1865. Mustered out June 9, 1865.

Niles, Nelson. Residence, Smithfield. 30. M. Operative. Enlisted Aug. 15, 1862. Mustered in Sept. 6, 1862. Teamster in quartermaster's department from Nov. 1862, until Feb. 1863. Died of dysentery at Smithfield, RI, Aug. 19, 1864. Interred at Intervale Cemetery, North Providence, RI.

Nolan, Patrick. Transferred from Co. D, Feb. 1, 1865. Mustered out June 9, 1865.

Northrup, Henry F. Residence, Portsmouth. 23. M. Farmer. Enlisted Aug. 16, 1862. Mustered in Sept. 6, 1862. Deserted at York, PA, Mar. 27, 1863.

Northrup, William H. Residence, Bristol. 23. M. Laborer. Enlisted Aug. 16, 1862. Mustered in Sept. 6, 1862. Wounded in action, shot in wrist, at Bethesda Church, VA, June 3, 1864. Wounded in action, shot in hand, at Petersburg, VA, June 22, 1864. Sick at Baltimore, MD, Jan. 1865 until mustered out May 14, 1865. Died July 18, 1903. Interred at Elm Grove Cemetery, North Kingstown, RI.

Olney, Obadiah. Residence, North Providence. 28. M. Stone cutter. Enlisted Aug. 19, 1862. Mustered in Sept. 6, 1862. Discharged for disability at Camp Banks, Alexandria, VA, Mar. 15, 1863. Died 1908. Interred at Pocasset Cemetery, Cranston, RI.

O'Donnell, John. Residence, Providence. 21. S. Harness maker. Enlisted Aug. 13, 1862. Mustered in Sept. 6, 1862. Wounded in action at Fredericksburg, VA, Dec. 13, 1862 and sent to hospital. Transferred to the Veteran Reserve Corps, Jan. 15, 1864. Interred at St. Mary's Cemetery, Pawtucket, RI.

O'Neil, Patrick. Transferred from Co. D, Feb. 1, 1865. Mustered out June 9, 1865.

Peckham, Benjamin. Residence, Bristol. 30. M. Farmer. Enlisted Aug. 16, 1862. Mustered in Sept. 6, 1862. Died of Yazoo Fever onboard the *David Tatum*, Aug. 11, 1863. Interred at Napoleonville, AR.

Pierce, Allen. Residence, Bristol. 19. M. Laborer. Enlisted Aug. 20, 1862. Mustered in Sept. 6, 1862. Mortally wounded in action at Cold Harbor, VA, June 6, 1864. Died of wounds at White House, VA, June 6, 1864. Interred at Yorktown National Cemetery, Grave 424. Cenotaph at North Burial Ground, Bristol, RI.

Potter, John. Residence, Bristol. 33. M. Farmer. Enlisted Aug. 15, 1862. Mustered in Sept. 6, 1862. Wounded in action, shot in back, at Fredericksburg, VA, Dec. 13, 1862. In ambulance corps from Jan. 1865, until June 1865. Mustered out June 9, 1965.

Price, James H. Residence, Newport. 22. S. Cork maker. Enlisted Aug. 3, 1862. Mustered in Sept. 6, 1862. Wounded in action, shot in jaw, at Fredericksburg, VA, Dec. 13, 1862, and sent to hospital. Transferred to Veteran Reserve Corps, Sept. 17, 1863. Died 1908. Interred at Pine Grove Cemetery, Coventry, RI.

Radekin, Edward A. Residence, Smithfield. 32. S. Laborer. Enlisted Aug. 12, 1862. Mustered in Sept. 6, 1862. Wounded in action at Fredericksburg, VA, Dec. 13, 1862. Transferred to Veteran Reserve Corps Sept. 1, 1863.

Reniers, Frank T. Residence, Bristol. 20. M. Farmer. Enlisted Aug. 20, 1862. Mustered in Sept. 6, 1862. Mustered out June 9, 1865.

Robinson, James. Residence, Newport. 32. M. Farmer. Enlisted Aug. 21, 1862. Mustered in Sept. Wounded in action at Fredericksburg, VA, Dec. 13, 1862. Borne as absent sick until Feb. 1863. Wounded in action at Spotsylvania Court House, VA, May 18, 1864. Brigade pioneer from Jan. 1865, until May 1865. Mustered out June 9, 1865. Died Aug. 16, 1875. Interred at Swan Point Cemetery, Providence, RI.

Seymour, Joseph R. Residence, Bristol. 24. M. Farmer. Enlisted Aug. 19, 1862. Mustered in Sept. 6, 1862. Mustered out June 9, 1865. Died 1918. Interred at South Burial Ground, Warren, RI.

Scannell, Dennis J. Residence, Worcester, MA. 21. S. Printer. Enlisted Aug. 19, 1862. Mustered in Sept. 6, 1862. Promoted to sergeant.

Sherman. Ezra H. Residence, Bristol. 19. S. Farmer. Enlisted Aug. 19, 1862. Mustered in Sept. 6, 1862. Wounded in action, shot in chest, at Fredericksburg, VA, Dec. 13, 1862. Wounded in action, shot in arm, at Bethesda Church, VA, June 3, 1864. Wounded in action, shot in chest, at Petersburg, VA, June 17, 1864. Mustered out June 9, 1865. Died June 21, 1909. Interred at North Burial Ground, Providence, RI.

Slater, James S. Residence, Smithfield. 21. S. Clerk. Enlisted Aug. 11, 1862. Mustered in Sept. 6, 1862. Discharged for disability at Pleasant Valley, MD, Oct. 25, 1862. Died Nov. 10, 1915. Interred at Slatersville Cemetery, North Smithfield, RI.

Spencer, George A. Residence, Bristol. 18. S. Farmer. Enlisted Aug. 16, 1862. Mustered in Sept. 6, 1862. Captured at Fredericksburg, VA, Dec. 13, 1862. Released and rejoined regiment Jan. 1863. Mustered out June 9, 1865. Died Sept. 19, 1914. Interred at North Burial Ground, Bristol, RI.

Spooner, Charles D. Residence, Newport. 31. M. Boat Builder. Enlisted Aug. 12, 1862. Mustered in Sept. 6, 1862. Promoted to corporal.

Stanfield, William. Transferred from Co. D, Feb. 1, 1865. Teamster from April until June. Mustered out June 9, 1865.

Sunderland, George B. Transferred from Co. D, Feb. 1, 1865. Mustered out June 9, 1865. Died Oct. 27, 1887. Interred at Wood River Cemetery, Richmond, RI.

Tanner, Richard D. Residence, Cranston. Enlisted and Mustered Jan. 7, 1865. Transferred to Co. G, June 9, 1865.

Taylor, Richard Edwin. Transferred from Co. D, Feb. 1, 1865. Mortally wounded in action, shot in face, at Petersburg, VA, April 2, 1865. Died of wounds at Washington, DC, April 16, 1865. Interred at Glenford Cemetery, Scituate, RI.

Thornton, Cyril P. Residence, Smithfield. 28. M. Mason. Enlisted Aug. 15, 1862. Mustered in Sept. 6, 1862. Teamster in quartermaster's department from Nov. 1862, until Jan. 1863. Absent sick from Jan. 1865, until June. Mustered out June 15, 1865. Died June 13, 1909. Interred at Brayton Cemetery, Warwick, RI.

Towle, John. Residence, Newport. 22. S. Printer. Enlisted Aug. 19, 1862. Wounded in action at Fredericksburg, VA, Dec. 13, 1862. Discharged for disability Mar. 17, 1863.

Utton, Samuel N. Residence, Newport. 37. M. Leather dyer. Enlisted Aug. 19, 1862. Mustered in Sept. 6, 1862. Wounded in action, shot in foot, at Spotsylvania Court House, VA, May 18, 1864, and sent to hospital. Mustered out June 15, 1865.

Walker, William H. Residence, East Providence. 34. M. Farmer. Enlisted Aug. 15, 1862. Mustered in Sept. 6, 1862. Promoted to corporal.

Wells, Perry J. Residence, Providence. 21. S. Blacksmith. Enlisted Aug. 12, 1862. Mustered in Sept 6, 1862. Absent sick at Pleasant Valley, MD, from Oct. 27, 1862, until Jan. 1863. Wounded in action, shot in leg, at the Crater, Petersburg, VA, July 30, 1864. Mustered out June 9, 1865. Died Mar. 12, 1896. Interred at Glenwood Cemetery, East Greenwich, RI.

Whitford, Clark. Residence, Bristol. 40. S. Shoemaker. Enlisted Aug. 16, 1862. Mustered in Sept. 6, 1862. Wounded in action at Fredericksburg, VA, Dec. 13, 1862. Wounded in action at Spotsylvania Court House, VA, May 18, 1864. Mustered out June 9, 1865. Died Nov. 26, 1899. Interred at North Burial Ground, Bristol, RI.

Whitman, Squire F. Transferred from Co. D, Feb. 1, 1865. Mustered out June 9, 1865. Died Jan. 7, 1883. Interred at Matteson-Whitman-Woodward Lot, West Greenwich Cemetery 46, West Greenwich, RI.

Whipple, Alfred H. Transferred from Co. D, Feb. 1, 1865. Mustered out June 9, 1865.

Winnesmann, Henry. Residence, Bristol. 23. M. Laborer. Enlisted Aug. 16, 1862. Mustered in Sept. 6, 1862. Wounded in action at Fredericksburg, VA, Dec. 13, 1862. Wounded in action at Bethesda Church, VA, June 3, 1864. Severely wounded, shot through both thighs at Petersburg, VA, June 16, 1864. Mustered out June 9, 1865.

Willis, Abel, Jr. Residence, Bristol. 32. M. Laborer. Enlisted Aug. 16, 1862. Mustered in Sept. 6, 1862. Mortally wounded in action at Fredericksburg, VA, Dec. 13, 1862. Died of wounds at Washington, DC, Dec. 28, 1862. Interred at Soldier's Home National Cemetery, Washington, DC. Grave 2209.

Wood, Frederick. Transferred from Co. D, Feb. 1, 1865. At division headquarters from Mar. until June 1865. Mustered out June 9, 1865.

Woodbury, John. Residence, Newport. 39. M. Pilot. Enlisted Aug. 19, 1862. Mustered in Sept. 6, 1862. Deserted in the face of the enemy at Fredericksburg, VA, Dec. 13, 1862.

COMPANY K

Captains

Durfee, George N. Residence, Tiverton. 18. S. Clerk. Commissioned Sept. 4, 1862. Mustered in Sept. 6, 1862. Wounded in action, Dec. 13, 1862 at Fredericksburg, VA. Resigned Mar. 20, 1863. Died Jan. 15, 1915. Interred at Briggs Lot, Tiverton Cemetery 65, Tiverton, RI.

Wilbur, George A. Promoted from first lieutenant Co. K, Mar. 1, 1863. Detailed on special duty as a member of general court-martial at division headquarters Dec. 2, 1864. Mustered out June 9, 1865. Died June 9, 1906. Interred at Union Cemetery, North Smithfield, RI.

First Lieutenants

Bates, Gustavus D. Promoted from second lieutenant Co. E, May 23, 1863. Discharged for disability Oct. 14, 1863. Again commissioned first lieutenant Nov. 14, 1863. Promoted to captain Co. E, July 25, 1864.

Groves, Joseph. Residence, Providence. 25. M. Plumber. Commissioned Sept. 4, 1862. Mustered in Sept. 6, 1862. Resigned Jan. 13, 1863. Died Jan. 17, 1880. Interred at St. Francis Cemetery, Pawtucket, RI.

Moore, Winthrop A. Promoted from second lieutenant Co. A, Jan. 9, 1864. Transferred to Co. D, Feb. 1, 1865.

Wilbur, George A. Promoted from second lieutenant Co. E, Jan. 13, 1863. Promoted to captain Co. K. Mar. 1, 1863.

Second Lieutenants

Healey, Charles T. Residence, Providence. 26. S. Engineer. Commissioned Sept. 6, 1862. Mustered in Sept. 6, 1862. Resigned

Jan. 7, 1863. Died July 12, 1885. Interred at Mt. Hope Cemetery, Boston, MA.

Perkins, Benjamin G. Promoted from sergeant Co. K, Mar. 1, 1863. Promoted to first lieutenant Co. A, July 1, 1863.

Sullivan, John Promoted from sergeant Co. D, Jan. 7, 1863. Promoted to adjutant Mar. 1, 1863.

Webb, William W. Transferred from Co. B, Dec. 28, 1863. Mustered out June 9, 1865. Died May 6, 1897. Interred at Grace Church Cemetery, Providence, RI.

First Sergeants

Bennett, George W. Residence, Foster. 24. S. Carpenter. Enlisted Aug. 14, 1862. Mustered in Sept. 6, 1862. Wounded in action, shot in right ankle, at Fredericksburg, VA, Dec. 13, 1862. Sent to hospital and never returned to regiment. Transferred to the Veteran Reserve Corps Oct. 31, 1863. Died May 6, 1881. Interred at Pine Grove Cemetery, Coventry, RI.

Robbins, Philander T. Promoted from sergeant Mar. 1, 1863. Mustered out June 9, 1865. Died Jan. 8, 1899. Interred at Togus National Cemetery. Grave 1469.

Sergeants

Belcher, Jonathan S. Residence, Smithfield. 27. M. Manufacturer. Enlisted Aug. 8, 1862. Mustered on Sept. 6, 1862. Promoted to second lieutenant 14th RI Heavy Artillery Jan. 15, 1864. Died in 1867 at New Orleans, LA. Interred at New Orleans, LA.

Colvin, Charles F. Promoted from private. Mustered out June 9, 1865. Died Nov. 19, 1879. Interred at Greenwood Cemetery, Coventry, RI.

Farnum, Samuel. Promoted from corporal June 4, 1863. Appointed captain in the 14th RI Heavy Artillery, and discharged to accept the appointment Dec. 22, 1863. Drowned on return from war Oct. 15, 1865. Cenotaph in Friends Burying Ground, Uxbridge, MA.

Fiske, Alfred. Transferred from Co. B, Feb. 1, 1865. Mustered out May 29, 1865. Died May 25, 1879. Interred at Oak Grove Cemetery, Pawtucket, RI.

Harrington, Stephen A. Residence, Scituate. 24. M. Peddler. Enlisted Aug. 8, 1862. Mustered in Sept. 6, 1862. Absent sick from Oct. 27, 1862 until Feb. 1863. Absent sick from Jan. 1865 until April 1865. Mustered out June 9, 1865. Died April 6, 1889. Interred at Jeremiah Harrington Lot, Scituate Cemetery 88, Scituate, RI.

Hill, Charles E. Promoted from private. Promoted to second lieutenant USCT Aug. 7, 1864. Died Oct. 9, 1895. Interred at Toulon Cemetery, Toulon, IL.

Hopkins, William D. Promoted from private. Died of dysentery at Providence, RI, Oct. 4, 1863. Interred at Swan Point Cemetery, Providence, RI.

Marsh, Peter A. Residence, Providence. 27. M. Moulder. Enlisted Aug. 8, 1862. Mustered in Sept. 6, 1862. Discharged for disability at Portsmouth Grove, RI, May 5, 1863.

Nottage, John S. Transferred from Co. B, Feb. 1, 1865. Discharged for disability at Portsmouth Grove, June 1,1865. Died Aug. 31, 1896. Interred at North Burial Ground, Providence, RI.

Perkins, Benjamin G. Residence, Warwick. 30. M. Wheelwright. Enlisted Aug. 8, 1862. Mustered in Sept. 6, 1862. Promoted to second lieutenant Co. K, Mar. 1, 1863.

Potter, George H. Promoted from corporal Mar. 1, 1865. Mustered out June 9, 1865. Died Jan. 28, 1908. Interred at Riverside Cemetery, Sterling, CT.

Richter, Henry M. Transferred from Co. G, Sept. 1, 1864. Promoted to sergeant major Nov. 4, 1864.

Robbins, Philander T. Promoted from corporal Jan. 1, 1863. Promoted to sergeant Mar. 1, 1863.

Corporals

Austin, John F. Residence, Scituate. 21. M. Carder. Enlisted Aug. 8, 1862. Mustered in Sept. 6, 1862. Wounded in action, shot in neck, at Fredericksburg, VA, Dec. 13, 1862. Discharged for disability at Portsmouth Grove, RI, Mar. 13, 1863. Died Dec. 17, 1886. Interred at Manchester Cemetery, Coventry, RI.

Griffin, Joseph H. Transferred from Co. H, Mar. 9, 1865. Mustered out June 9, 1865. Interred at Riverbend Cemetery, Westerly, RI.

Farnum, Samuel. Promoted from corporal. Promoted to sergeant June 4, 1863.

Hackett, John. Promoted from private. Mustered out June 9, 1865. Died May 14, 1867. Interred at Swan Point Cemetery, Providence, RI.

Hackett, Patrick. Residence, Providence. 23. S. Peddler. Enlisted Aug. 12, 1862. Mustered in Sept. 6, 1862. Transferred to the Veteran Reserve Corps, Aug 6, 1864. Served in Color Guard. Died Sept. 24, 1878. Interred at State Farm Cemetery, Cranston, RI.

Howland, Franklin. Residence, Providence. 40. M. Cooper. Enlisted Aug. 8, 1862. Mustered in Sept. 6, 1862. Discharged for disability from U. S. General Hospital, Cincinnati, OH, Aug. 17, 1864.

Nye, Charles P. Promoted from private. Wounded in action, shot in thigh, at Petersburg, VA, July 8, 1864. Mustered out June 9, 1865. Died Jan. 31, 1922. Interred at North Burial Ground, Providence, RI.

Nye, Isaac. Promoted from private. Mortally wounded in action at Spotsylvania Court House, VA, May 18, 1864. Died of wounds in hospital at Alexandria, VA, May 30, 1864. Served in Color Guard.

Interred at Alexandria National Cemetery. Grave 1985. Cenotaph in Manchester Cemetery, Coventry, RI.

Potter, George H. Promoted from private June 4, 1863. Wounded in action, shot in hand, at Cold Harbor, VA, June 8, 1864. Wounded in action at Poplar Spring Church, VA, Sept. 30, 1864. Promoted to sergeant Mar. 1, 1865.

Potter, Roswell H. Residence, Providence. 25. M. Laborer. Enlisted Aug. 8, 1862. Mustered in Sept. 6, 1862. Absent sick from Nov. 1862, until Jan. 1863. Died of Yazoo Fever at Milldale, MS, July 22, 1863. Interred at Vicksburg National Cemetery. Section G, Grave 5636.

Reynolds, Edward S. Promoted from private. Killed in action near Cold Harbor, VA, June 2, 1864. Interred at Cold Harbor National Cemetery.

Robbins, Philander T. Residence, Foster. 21. S. Farmer. Enlisted July 8, 1862. Mustered in Sept. 6, 1862. Promoted to sergeant Jan. 1, 1863.

Smith, George H. Residence, Scituate. 23. S. Jeweler. Enlisted Aug. 11, 1862. Mustered in Sept. 6, 1862. Died of typhoid at Falmouth, VA, Jan. 3, 1863. Interred at Clayville Cemetery, Foster, RI.

Vance, Fleming Residence, Providence. 22. S. Moulder. Enlisted Aug. 8, 1862. Mustered in Sept. 6, 1862. Mustered out June 9, 1865.

Young, George W. Residence, Providence. 27. M. Baker. Enlisted Aug. 8, 1862. Mustered in Sept. 6, 1862. Discharged for disability at Portsmouth Grove, RI, Mar. 16, 1864. Died March 24, 1891. Interred at West Greenwich Cemetery 43, West Greenwich, RI.

Musician

Abbott, William A. Residence, Providence. 35. M. Moulder. Enlisted Aug. 8, 1862. Mustered in Sept. 6, 1862. Mustered out

June 9, 1865. Died Sept. 20, 1896. Interred at Moshassuck Cemetery, Central Falls, RI.

Wagoner

Austin, John A. Residence, Smithfield. Enlisted Aug. 8, 1862. Mustered in Sept. 6, 1862. On extra duty in quartermaster's department as teamster from Nov. 1862, until Feb. 1863. In quartermaster's department from April 1865, to June 1865. Mustered out June 9, 1865.

Privates

Ashworth, William. Residence, Warwick. 18. S. Farmer. Enlisted Aug. 15, 1862. Mustered in Sept. 6, 1862. Died of typhoid in hospital at Lexington, KY, Jan. 30, 1864. Interred at St. Phillips Episcopal Cemetery, West Warwick, RI.

Aylesworth, Albert H. Residence, Scituate. 19. M. Farmer. Enlisted Aug. 14, 1862. Mustered in Sept. 6, 1862. Absent sick at City Point, VA, Jan. 1865, and borne as absent sick until mustered out May 31, 1865. Died Jan. 30, 1903. Interred at North Burial Ground, Providence, RI.

Bateman, George. Residence, Providence. 30. M. Spinner. Enlisted Aug. 14, 1862. Mustered in Sept. 6, 1862. On detached service in hospital from Nov. 1862, until Jan. 1863. Died of Yazoo Fever at Covington, KY Aug. 20, 1863. Interred at Camp Nelson National Cemetery. Section G, Grave 2023.

Battey, Hiram S. Residence, Johnston. 18. S. Farmer. Enlisted Aug. 8, 1862. Mustered in Sept. 6, 1862. Detached to Battery E, Second U.S. Artillery. Died of dysentery at Marine Hospital, Cincinnati, OH, Aug. 16, 1863. Interred at Spring Grove National Cemetery, Cincinnati, OH. Grave 521. Cenotaph at Atwood-Salisbury Lot, Scituate Cemetery 44, Scituate, RI.

Bigelow, Edward. Residence, Woonsocket. Enlisted Aug. 17, 1862. Mustered in Sept. 6, 1862. Wounded in action, May 15, 1864 at Spotsylvania, VA. Mustered out June 9, 1865.

Briggs, James A. Residence, Foster. 20, S, Farmer. Enlisted Aug. 18, 1862. Mustered in Sept. 6, 1862. Discharged for disability at West's Building Hospital, Baltimore, MD, Jan 1, 1863. Died Sept. 9, 1896. Interred at Sand Hill Cemetery, Foster, RI.

Brown, John D. Residence, Scituate. 29, M. Farmer. Enlisted Aug. 8, 1862. Mustered in Sept. 6, 1862. Transferred to Veteran Reserve Corps, Jan. 15, 1864. Died 1917. Interred at Oak Grove Cemetery, Pawtucket, RI.

Bryden, Wilson C. Residence, Burrillville. 18. S. Broom maker. Enlisted Aug. 6, 1862. Mustered in Sept. 6, 1862. Discharged for disability at Philadelphia, PA, Nov. 14, 1862. Died of typhoid fever contracted in the service Dec. 29, 1862 at Webster, MA. Interred at Mount Zion Cemetery, Webster, MA.

Bunn, James A. Residence, Glocester. 18. S. Farmer. Enlisted Aug. 18, 1862. Mustered in Sept. 6, 1862. Mustered out June 9, 1865.

Clarke, Stephen A. Residence, Hopkinton. 29. M. Farmer. Enlisted Aug. 25, 1862. Mustered in Sept. 6, 1862. Killed in action at Poplar Spring Church, VA, Sept. 30, 1864. Interred at Poplar Grove National Cemetery. Grave 3171. Cenotaph in Wood River Cemetery, Richmond, RI.

Collins, Charles H. Residence, North Mansfield, MA. 20, S. Spinner. Enlisted, July 18, 1862. Mustered in Sept. 6, 1862. Wounded in action, shot in thigh, at Petersburg, VA, June 26, 1864. Mustered out June 9, 1865. Died Oct. 29, 1892. Interred at North Burial Ground, Providence, RI.

Collins, Nehemiah R. Residence, Scituate. 44. M. Farmer. Enlisted Aug. 8, 1862. Mustered in Sept. 6, 1862. Wounded in action, shot in head, at Fredericksburg, VA, Dec. 13, 1862. Discharged for disability at Portsmouth Grove, RI, Feb. 2, 1864. Died Aug. 9, 1888. Interred at Collins Lot, Scituate Cemetery 113, Scituate, RI.

Colvin, Charles F. Residence, Scituate. 26. M. Farmer. Enlisted Aug. 8, 1862. Mustered in Sept. 6, 1862. Promoted to sergeant June 1, 1863.

Cole, John H. Residence, Scituate. 34. M. Farmer. Enlisted Aug. 8, 1862. Mustered in Sept. 6, 1862. Mustered out June 9, 1865. Died 1896. Interred at Knotty Oak Cemetery, Coventry, RI.

Cole, Henry S. Residence, Foster. 33. M. Farmer. Enlisted Aug. 16, 1862. Mustered in Sept. 6, 1862. Killed in action at Fredericksburg, VA, Dec. 13, 1862. Cenotaph in Cole Lot, Foster Cemetery 52, Foster, RI.

Colwell, George. Residence, Johnston. 27. S. Farmer. Enlisted Aug. 8, 1862. Mustered in Sept. 6, 1862. Mustered out June 9, 1865. Died July 12, 1912. Interred at Knotty Oak Cemetery, Coventry, RI.

Corbin, Amasa N. Residence, Scituate. 42. M. Farmer. Enlisted Aug. 19, 1862. Mustered in Sept. 6, 1862. Died of typhoid at Falmouth, VA, Dec. 24, 1862. Interred at Clayville Cemetery, Foster, RI.

Corbin, William H. Residence, Scituate. 18. S. Operator. Enlisted Aug. 19, 1862. Mustered in Sept. 6, 1862. Wounded in action, shot in head, at Cold Harbor, VA, June 8, 1864. Wounded in action, shot in head, at Petersburg, VA, June 16, 1864. Mustered out June 9, 1865. Died Dec. 26, 1922. Interred at Oak Grove Cemetery, Pawtucket, RI.

Cornell, Ira. Residence, Coventry. 44. M. Farmer. Enlisted Aug. 14, 1862. Mustered in Sept. 6, 1862. Wounded in action at Fredericksburg, VA, Dec. 13, 1862. Borne on the rolls of the Rhode Island Adjutant General as having deserted at Cincinnati, OH, April 1, 1865. According to the vital records of Coventry, RI, he died at Cincinnati, OH, April 1, 1863. "Drowned in the Ohio River in the attempt of crossing it in the line of duty." Carried on this roll as having drowned April 1, 1863 at Cincinnati, OH. Cenotaph in Pine Grove Cemetery, Coventry, RI.

Cornell, Ira, Jr. Residence, Coventry. 18. S. Farmer. Enlisted Aug. 15, 1862. Mustered in Sept. 6, 1862. Discharged for disability at Portsmouth Grove, RI, Oct. 14, 1864. Died of disease contracted in the service at Coventry, RI April 27, 1867. Interred at Pine Grove Cemetery, Coventry, RI.

Corey, Charles H. Residence, North Providence. 20. S. Spinner. Enlisted Aug. 8, 1862. Mustered in Sept. 6, 1862. Died of dysentery at Camp Dennison, OH, Sept. 15, 1863. Interred at Spring Grove National Cemetery. Cincinnati, OH. Section 21, Grave 837.

Cummings, Chester C. Residence, Foster. 31. M. Farmer. Enlisted Aug. 8, 1862. Mustered in Sept. 6, 1862. Transferred to Veteran Reserve Corps, Oct. 31, 1863. Died 1874. Interred at North Burial Ground, Providence, RI.

Earle, Albert. Residence, Scituate. 40. M. Manufacturer. Enlisted Aug. 14, 1862. Mustered in Sept. 6, 1862. Wounded in action, shot in face, at Fredericksburg, VA, Dec. 13, 1862. Transferred to the Veteran Reserve Corps Sept. 1, 1863.

Farnum, Samuel. Residence, Uxbridge, MA. 22. S. Teacher. Enlisted Aug. 8, 1862. Mustered in Sept. 6, 1862. Promoted to corporal Jan. 4, 1863.

Farrow, Enos. Residence, Foster. 32. M. Farmer. Enlisted Aug. 16, 1862. Mustered in Sept. 6, 1862. Died of typhoid at Washington, DC, Dec. 3, 1862. Interred at Soldier's Home National Cemetery. Section H, Grave 3552.

Field, George A. Residence, Scituate. 19. S. Farmer. Enlisted Aug. 8, 1862. Mustered in Sept. 6, 1862. Absent sick in hospital from Oct. 27, 1862, until Nov. 9, 1862. Sick in hospital from Aug. 28, 1863, until Sept. 17, 1863. Died of dysentery at general hospital, Lexington, KY, April 5, 1864. Interred at Isaac Field Lot, Scituate Cemetery 59, Scituate, RI.

Fleming, David J. Residence, North Providence. Enlisted and Mustered in April 6, 1865. Transferred to Co. G, June 9, 1865.

Gavitt, James W. Residence, Coventry. 27. S. Farmer. Enlisted Aug. 14, 1862. Mustered in Sept. 6, 1862. Wounded in action, shot in hand, at Petersburg, VA, June 16, 1864. Wounded in action at Poplar Spring Church, VA, Sept. 30, 1864. Mustered out June 9, 1865. Died Jan. 19, 1895. Interred at Manchester Cemetery, Coventry, RI.

Greene, Lewis E. Residence, Scituate. 20. M. Farmer. Enlisted Aug. 8, 1862. Mustered in Sept. 6, 1862. Discharged for disability from West's Building Hospital, Baltimore, MD, June 20, 1863. Died June 4, 1910. Interred at Acotes Hill Cemetery, Glocester, RI.

Hackett, John. Residence, Providence. 25. S. Moulder. Enlisted Aug. 8, 1862. Mustered in Sept. 6, 1862. Promoted to corporal June 4, 1863.

Harkness, Henry A. Residence, Coventry. 33. S. Carpenter. Enlisted Aug. 14, 1862. Mustered in Sept. 6, 1862. Wounded in action, shot in stomach, at Petersburg, VA, June 26, 1864, and sent to hospital. Discharged for disability Jan. 26, 1865. Died Jan. 15, 1880. Interred at Manchester Cemetery, Coventry, RI.

Harrington, William O. Residence, Foster. 32. M. Farmer. Enlisted Aug. 14, 1862. Mustered in Sept. 6, 1862. Mustered out June 9, 1865. Died Mar. 18, 1904. Interred at Moosup Valley Cemetery, Foster, RI.

Hawkins, George W. Residence, Scituate. 30. M. Spinner. Enlisted Aug. 14, 1862. Mustered in Sept. 6, 1862. Mustered out June 9, 1865. Died Aug. 5, 1909. Interred at Rockland Cemetery, Scituate, RI.

Hill, Charles E. Residence, Providence. 27. M. Gilder. Enlisted Aug. 18, 1862. Mustered in Sept. 6, 1862. Promoted to sergeant.

Holloway, Thomas T. Residence, Foster. 26. M. Farmer. Enlisted Aug. 16, 1862. Mustered in Sept. 6, 1862. Died of Yazoo Fever at Union Hospital, Memphis, TN, Aug. 23, 1863. Interred at Keech-Winsor Lot, Foster Cemetery 17, Foster, RI.

Hopkins, Adoniram J. Residence, Foster. 21. S. Farmer. Enlisted Aug. 12, 1862. Mustered in Sept. 6, 1862. Discharged for disability at Baltimore, MD, Mar. 4, 1863. Died April 7, 1930. Interred at Hopkins Mills Cemetery, Foster, RI.

Hopkins, Ashael A. Residence, Foster. 22. S. Farmer. Enlisted Aug. 9, 1862. Mustered in Sept. 6, 1862. Died of dysentery at Loudon, TN, April 11, 1864. Interred at Tucker Hollow Cemetery, Foster, RI.

Hopkins, Darius A. Residence, Scituate. 38. S. Farmer. Enlisted Aug. 16, 1862. Mustered in Sept. 6, 1862. Died of Yazoo Fever at Cincinnati, OH, Sept. 29, 1863. Interred at Clayville Cemetery, Foster, RI.

Hopkins, John. Residence, Foster. 29. S. Farmer. Enlisted Aug. 16, 1862. Mustered in Sept. 6, 1862. Died of typhoid at regimental hospital at Newport News, VA, Mar. 1, 1863. Interred at Hopkins Lot, Foster Cemetery 74, Foster, RI.

Hopkins, John E. Residence, Foster. 25. M. Farmer. Enlisted Aug. 12, 1862. Mustered in Sept. 6, 1862. Died of dysentery at Memphis, TN, Aug. 17, 1863. Interred at Hopkins Mills Cemetery, Foster, RI.

Hopkins, William D. Residence, Scituate. 33. S. Moulder. Enlisted Aug. 9, 1862. Mustered in Sept. 6, 1862. Promoted to sergeant June 8, 1863.

Jordan, John F. Residence, Scituate. 25. M. Dresser. Enlisted Aug. 8, 1862. Mustered in Sept. 6, 1862. On detached service at division headquarters Jan. 1863. Mustered out June 9, 1865. Died Jan. 14, 1905. Interred at Glenford Cemetery, Scituate, RI.

Jordan, William H. Residence, Coventry. 22. S. Farmer. Enlisted Aug. 11, 1862. Mustered in Sept. 6, 1862. In quartermaster's department from Jan. 1865 until June 1865. Mustered out June 9, 1865. Died Aug. 1, 1925. Interred at Hopkins Hollow Cemetery, Coventry, RI.

Keach, Henry M. Residence, Blackstone, MA. 18. S. Farmer. Enlisted June 27, 1862. Mustered in Sept. 6, 1862. Deserted at Pleasant Valley, MD, Oct. 10, 1862.

Kelley, Michael R. Residence, Scituate. 35. M. Laborer. Enlisted Aug. 8, 1862. Mustered in Sept. 6, 1862. Mustered out June 9, 1865. Died Sept. 4, 1870. Interred at St. Patrick's Cemetery, Providence, RI.

Kenny, Steakley. Residence, Richmond. 23. S. Farmer. Enlisted Aug. 26, 1862. Mustered in Sept. 6, 1862. On detached service in ambulance corps from Nov. 1862, until Feb. 1863. Mustered out June 9, 1865.

Kenyon, Abel B. Residence, Hopkinton. 25. M. Farmer. Enlisted Aug. 25, 1862. Mustered in Sept. 6, 1862. Wounded in action, shot in head, at Jackson, MS, July 13, 1863. Wounded in action, shot in hand, at Spotsylvania Court House, VA, May 12, 1864. Absent sick at Philadelphia, Jan. 1865. Mustered out June 9, 1865. Died Nov. 28, 1911. Interred at Rockville Cemetery, Hopkinton, RI.

Knight, Jeremiah F. Residence, West Greenwich. 26. M. Farmer. Enlisted Aug. 8, 1862. Mustered in Sept. 6, 1862. On ambulance corps from Nov. 1862, until Feb. 1863, and from Jan. 1865, until June 1865. Mustered out June 9, 1865. Died Sept. 23, 1903. Interred at Knotty Oak Cemetery, Coventry, RI.

Lewis, Edward S. Residence, Scituate. 19. S. Farmer. Enlisted Aug. 8, 1862. Mustered in Sept. 6, 1862. Wounded in action at Spotsylvania Court House, VA, May 18, 1864. Absent sick at Portsmouth Grove, RI, from Jan. 1865, until mustered out June 14, 1865. Died Dec. 29, 1920. Interred at Brayton Cemetery, Warwick, RI.

Lillibridge, Charles P. Residence, Scituate. 40. S. Shoemaker. Enlisted Aug. 8, 1862. Mustered in Sept. 6, 1862. Discharged for disability at Newport News, VA, Mar. 19, 1863. Died April 20, 1889. Interred at Lillibridge Lot, Exeter Cemetery 6, Exeter, RI.

Maxon, Joel C. Residence, Hopkinton. 33. M. Farmer. Enlisted Aug. 25, 1862. Mustered in Sept. 6, 1862. Absent sick from Nov. 17, 1862, until Feb. 1863. Discharged for disability at Louisville, KY, Aug. 13, 1863. Died of dysentery Sept. 24, 1863 at Hopkinton, RI. Interred at Oak Grove Cemetery, Hopkinton, RI.

Nye, Charles P. Residence, Richmond. 22. S. Clerk. Enlisted Aug. 26, 1862. Mustered in Sept. 6, 1862. Promoted to corporal April 4, 1863.

Nye, Isaac. Residence, Coventry. 25. S. Carpenter. Enlisted Aug. 14, 1862. Mustered in Sept. 6, 1862. Promoted to corporal.

Parker, Joseph. Residence, Cranston. 32, S. Laborer. Enlisted Aug. 8, 1862. Mustered in Sept. 6, 1862. Wounded in action, shot in hand, at Spotsylvania Court House, VA, May 12, 1864. Mustered out June 9, 1865.

Perry, Joseph B. Residence, Richmond. 26. M. Farmer. Enlisted Aug. 26, 1862. Mustered in Sept. 6, 1862. Discharged for disability Jan. 27, 1863. Died Sept. 16, 1908. Interred at Pine Grove Cemetery, Hopkinton, RI.

Pierce, Benjamin S. Residence, Scituate. 18. S. Farmer. Enlisted Aug. 11, 1862. Mustered in Sept. 6, 1862. Mustered out June 9, 1865. Died Jan. 22, 1907. Interred at Clayville Cemetery, Foster, RI.

Pierce, Edwin O. Residence, Scituate. 18. S. Spinner. Enlisted Aug. 8, 1862. Mustered in Sept. 6, 1862. Transferred to Veteran Reserve Corps, Jan. 28, 1865. Died 1924. Interred at West Warwick Cemetery 26, West Warwick, RI.

Potter, George H. Residence, Foster. 34. S. Blacksmith. Enlisted Aug. 14, 1862. Mustered in Sept. 6, 1862. Wounded in action, shot in shoulder, at Fredericksburg, VA, Dec. 13, 1862. Promoted to corporal June 4, 1863.

Potter, Pardon Knight. Residence, Cranston. 28. M. Stage Driver. Enlisted Aug. 4, 1862. Mustered in Sept. 6, 1862. Left sick at

Pleasant Valley, MD, Oct. 27, 1862. Discharged for disability at Portsmouth Grove, RI, Mar. 24, 1863. Died Jan. 10, 1913. Interred at Cottrell Cemetery, Scituate, RI.

Prey, Esais. Residence, Foster. 43. M. Farmer. Enlisted Aug. 14, 1862. Mustered in Sept. 6, 1862. Wounded in action, shot in jaw, at Spotsylvania Court House, VA, May 12, 1864, and sent to General Hospital. Discharged for disability at Portsmouth Grove, RI, Nov. 17, 1864. Died Jan. 31, 1911. Interred at Munson Cemetery, Putnam, CT.

Reynolds, Edward S. Residence, Scituate. 18. S. Carder. Enlisted Aug. 14, 1862. Mustered in Sept. 6, 1862. Promoted to corporal April 4, 1863.

Roberts, Henry A. Residence, Warwick. 34. M. Farmer. Enlisted July 1862. Mustered in Sept. 6, 1862. On extra duty in quartermaster's department from Nov. 1862, until Feb. 1863. Wounded in action, shot in arm, at Petersburg, VA, June 29, 1864. Mustered out June 9, 1865. Interred at Roberts-Matthewson Lot, Cranston Cemetery 16, Cranston, RI.

Rounds, Chester P. Residence, Foster. 22. S. Farmer. Enlisted Aug. 16, 1862. Mustered in Sept. 6, 1862. Wounded in action, shot in hand, at Spotsylvania Court House, VA, May 12, 1864. Absent sick at Portsmouth Grove, RI, from Jan. 1, 1865, until Mar. 1865, when he was transferred to Veteran Reserve Corps. Died 1925. Interred at Pocasset Cemetery, Cranston, RI.

Salisbury, Alpheus. Residence, Scituate. 30. M. Weaver. Enlisted Aug. 8, 1862. Mustered in Sept. 6, 1862. Mortally wounded in action, shot in neck, at Fredericksburg, VA, Dec. 13, 1862. Discharged for disability Feb. 3, 1863. Died of wounds July 4, 1863 at Scituate, RI. Interred at Clayville Cemetery, Foster, RI.

Searle, Benjamin F. Residence, Cranston. 47. M. Clothier. Enlisted Aug. 5, 1862. Mustered in Sept. 6, 1862. Left sick at Baltimore, MD, Mar. 26, 1863. Discharged for disability at Cincinnati, OH, Dec. 30, 1863. Died Aug. 10, 1896. Interred at Freeborn Brayton Cemetery, Cranston, RI.

Searle, Henry E. Residence, Scituate. 18. S. Weaver. Enlisted Aug. 8, 1862. Mustered in Sept. 6, 1862. Wounded in action, shot in hand, May 12, 1864 at Spotsylvania Court House, Virginia. Transferred to the Veteran Reserve Corps Jan. 1, 1865.

Shippee, Horace J. Residence, Foster. 42. M. Farmer. Enlisted Aug. 14, 1862. Mustered in Sept. 6, 1862. On detached service at division headquarters from Dec. 1862, until Feb. 1863. In Ninth Corps commissary department from Jan. 1865, until May 1865. Mustered out June 9, 1865. Died Nov. 15, 1894. Interred at Elm Grove Cemetery, North Kingstown, RI.

Simmons, George. Residence, Foster. 40. M. Farmer. Enlisted Aug. 14, 1862. Mustered in Sept. 6, 1862. Wounded in action, in head, at Fredericksburg, VA, Dec. 13, 1862. Sent to Portsmouth Grove, RI, and borne as absent sick until Feb. 1863. Killed in action at Spotsylvania Court House, VA, May 12, 1864. Cenotaph at Walker-Randall Lot, Foster, Foster Cemetery 54, Foster, RI.

Simmons, Isaac. Residence, Foster. 21. S. Farmer. Enlisted Aug. 14, 1862. Mustered in Sept. 6, 1862. Mustered out June 9, 1865. Interred at Pocasset Cemetery, Cranston, RI.

Simpson, John. Residence, Smithfield. 23. M. Laborer. Enlisted Aug. 6, 1862. Mustered in Sept. 6, 1862. Wounded in action at Fredericksburg, VA, Dec. 13, 1862. Deserted at Falmouth, VA, Jan. 24, 1863.

Smith, James T. Residence, North Providence. 24. M. Carder. Enlisted Aug. 8, 1862. Mustered in Sept. 6, 1862. Mustered out June 9, 1865. Died April 2, 1913. Interred at First Cemetery, East Greenwich, RI.

Smith, Joseph. Residence, North Providence. 22, S. Spinner. Enlisted Nov. 3, 1863. Wounded in action, shot in abdomen at Petersburg, VA, June 27, 1864. Mustered out July 13, 1865. Died 1914. Interred at Pocasset Cemetery, Cranston, RI.

Studley, John N. Residence, Scituate. 26. M. Operator. Enlisted Aug. 14, 1862. Mustered in Sept. 6, 1862. Wounded in action, shot in knee, at Fredericksburg, VA, Dec. 13, 1862, and sent to hospital. Discharged for disability at Providence, April 3, 1863. Died March 13, 1879. Interred at Irwin-Hines Lot, Coventry Cemetery 81, Coventry, RI.

Taylor, James J. Residence, Smithfield. 28. M. Laborer. Enlisted Aug. 15, 1862. Mustered in Sept. 6, 1862. Mortally wounded in action at Bethesda Church, VA, June 3, 1864. Died of wounds at Finley Hospital, Washington, DC, July 6, 1864. Interred at Arlington National Cemetery. Section 13, Grave 5426.

Thurston, Caleb. Residence, Richmond. 28. M. Farmer. Enlisted Aug. 23, 1862. Mustered in Sept. 6, 1862. Left sick at Covington, KY, Aug. 23, 1863. Discharged for disability at Portsmouth Grove, RI, Feb. 29, 1864. Died May 15, 1883. Interred at Wood River Cemetery, Richmond, RI.

Waterman, Albert G. Residence, Cranston. 27. M. Peddler. Enlisted Aug. 11, 1862. Mustered in Sept. 6, 1862. Mustered out June 9, 1865. Died May 8, 1897. Interred at Resolved Waterman Lot, Johnston Cemetery 15, Johnston, RI.

Whitting, Hassan O. Residence, Smithfield. 23. S. Teacher. Mustered in Sept. 6, 1862. Transferred to Veteran Reserve Corps Mar. 16, 1864.

Williams, Olney D. Residence, North Providence. 22. S. Stone cutter. Enlisted Aug. 9, 1862. Mustered in Sept. 6, 1862. Killed in action at Fredericksburg, VA, Dec. 13, 1862.

Winsor, Albert A. Residence, Foster. 18. S. Farmer. Enlisted Aug. 16, 1862. Mustered in Sept. 6, 1862. Killed in action Fredericksburg, VA, Dec. 13, 1862. Cenotaph in Keech-Winsor Lot, Foster Cemetery 17, Foster, RI.

Wood, Oliver. Residence, Foster. 41. S. Carpenter. Enlisted Aug. 14, 1862. Mustered in Sept. 6, 1862. Mortally wounded in action at Bethesda Church, VA, June 3, 1864. Died of wounds at Cold Harbor, VA, June 5, 1864. Interred at Cold Harbor National Cemetery. Grave 796. Cenotaph in Line Cemetery, Foster, RI.

Weigand, Frederick. Residence, Providence. 40. S. Laborer. Enlisted Aug. 24, 1862. Appointed Regimental Color Sergeant Oct. 13, 1862. Promoted to second lieutenant Co. G, Jan. 7, 1863.

Young, Searles B. Residence, Foster. 21. S. Farmer. Enlisted Aug. 14, 1862. Mustered in Sept. 6, 1862. Wounded in action, shot in jaw and arm, at Fredericksburg, VA, Dec. 13, 1862. Discharged for disability at Washington, DC, Feb. 4, 1863. Died Aug. 12, 1925. Interred at North Foster Cemetery, Foster, RI.

COMPANY B (New Organization)

Captains

Remington, Daniel S. Promoted from first lieutenant Co. G, June 8, 1865. Mustered out July 13, 1865. Died May 6, 1912. Interred at North Burial Ground, Providence, RI.

Reynolds, William J. Residence, North Kingstown. 30. M. Jeweler. Enlisted Sept. 3, 1861. Resigned May 15, 1865. Died Aug. 4, 1913. Interred at Elm Grove Cemetery, North Kingstown, RI.

First Lieutenant

McKay, John, Jr. Transferred from Co. H, Feb. 1, 1865. Mustered out June 9, 1865. Died 1913. Interred at Greenwood Cemetery, Coventry, RI.

Second Lieutenant

Costello, George B. Residence, Providence. 21. S. File cutter. Enlisted in Co. C, 4th RI Vols. Sept. 10, 1861. Wounded in action at Petersburg, VA, Mar. 20, 1865. Promoted to first lieutenant Co. D, June 27, 1865.

First Sergeant

Collins, Albert R. Residence, Providence. 27. M. Painter. Enlisted in Co. B, 4th RI Vols. Sept. 5, 1861. Mustered out July 13, 1865. Died June 8, 1898. Interred at Spring Brook Cemetery, Mansfield, MA.

Sergeants

Burbank, James H. Residence, Providence. 23. S. Seaman. Enlisted in Co. K, 4th RI Vols. Sept. 25, 1861. Mustered out July

13, 1865. Died Feb. 15, 1911. Interred at Miltonvale Cemetery, Miltonvale, KS.

Burrill, Leroy S. Promoted from corporal July 1, 1865. Mustered out July 13, 1865. Died June 14, 1896. Interred at South Easton Cemetery, Easton, MA.

Jillson, Andrew. Residence, Cumberland. 23. S. Laborer. Enlisted in Co. E, 4th RI Vols. Sept. 10, 1861. Mustered out July 13, 1865. Died 1907. Interred at Wilcox Cemetery, Bellingham, MA.

Corporals

Allen, George H. Residence, Providence. 24. M. Baker. Enlisted in Co. B, 4th RI Vols. Sept. 6, 1861. Mustered out July 13, 1865. Died Jan. 26, 1915. Interred at North Burial Ground, Providence, RI.

Ballou, Welcome. Residence, Burrillville. 39. M. Farmer. Enlisted in Co. E, 4th RI Vols. Sept. 10, 1861. Absent on furlough June 1865. Mustered out July 13, 1865. Died Feb. 16, 1904. Interred at Ballou-Buffum Lot, Burrillville Cemetery 37, Burrillville, RI.

Burlingame, Benjamin W. Residence, Warwick. 38, M. Enlisted in Co. C, 4th RI Vols. Aug. 7, 1862. Mustered out June 9, 1865. Died April 25, 1896. Interred at Knotty Oak Cemetery, Coventry, RI.

Burrill, Leroy S. Residence, Norwich, CT. 21. S. Shoemaker. Enlisted in Co. K, 4th RI Vols. Sept. 25, 1861. Promoted to sergeant July 1, 1865.

Clancy, Thomas. Residence, Smithfield. 21. S. Laborer. Enlisted in Co. C, 4th RI Vols. Sept. 10, 1861. Mustered out July 13, 1865.

Clough, Charles F. Residence, Cumberland. 21. S. Laborer. Enlisted in Co. E, 4th RI Vols. Sept. 10. 1861. Mustered out July 13, 1865. Died Dec. 28, 1927. Interred at Greenwood Memorial Terrace, Spokane, WA.

Coggeshall, Thomas J.L. Residence, Warwick. 18. S. Farmer. Enlisted in Co. K, 4th RI Vols. Sept. 23, 1861. Mustered out July 13, 1865.

Donnelly, James. Promoted from private. Mustered out July 13, 1865.

Griffin, Sylvester. Residence, Smithfield. 21. S. Operative. Enlisted in Co. E, 4th RI Vols. Sept. 10, 1861. Mustered out July 13, 1865. Died Dec. 19, 1901. Interred at St. Johns Cemetery, Worcester, MA.

Murray, John J. Promoted from private. Mustered out July 13, 1865.

Musicians

Boyle, John J. Residence, Cumberland. 21. S. Operative. Enlisted in Co. E, 4th RI Vols. Sept. 10, 1861. Mustered out July 13, 1865.

Smith, Matthew. Residence, Providence. 26. S. Musician. Enlisted in Co. H, 4th RI Vols. Oct. 1, 1861. Mustered out July 13, 1865.

Privates

Annes, Jesse L. Residence, Providence. 19. S. Butcher. Enlisted in Co. C, 4th RI Vols. Sept. 5, 1861. Mustered out July 13, 1865. Died July 11, 1870. Interred at North Burial Ground, Providence, RI.

Anthony, Samuel H. Residence, Providence. 18. S. Clerk. Enlisted in Co. E, 4th RI Vols. Aug. 30, 1862. Mustered out June 21, 1865. Died Dec. 4, 1881. Interred at Swan Point Cemetery, Providence, RI.

Armes, Nicholas B. Residence, Providence. Enlisted in Co. B, 4th RI Vols. Aug. 15, 1862. Mustered out June 9, 1865.

Arnold, Benjamin F. Transferred from Co. C, June 9, 1865. Mustered out July 13, 1865. Drowned in New York Harbor, July 15, 1865. Interred at Elm Grove Cemetery, North Kingstown, RI.

Ballou, George E. Residence, Burrillville. 31. S. Farmer. Enlisted in Co. D, 4th RI Vols. Sept. 23, 1861. Died of typhoid at Lincoln Hospital, Washington, DC, Jan. 27, 1865. Interred at Arlington National Cemetery. Section 13, Grave 9396.

Barry, David. Transferred from Co. B, June 9, 1865. Mustered out July 13, 1865.

Bates, George E. Residence, Providence. 22. M. Jeweler. Enlisted in Co. B, 4th RI Vols. Sept. 5, 1861. Mustered out June 9, 1865. Died April 27, 1890. Interred at Swan Point Cemetery, Providence, RI.

Blake, James. Residence, Providence. 37. M. Gardiner. Enlisted in Co. K, 4th RI Vols. Sept. 20, 1861. Mustered out July 13, 1865. Died 1900. Interred at Pascoag Cemetery, Burrillville, RI.

Bligh, John. Residence, Coventry. 17. S. Weaver. Enlisted in Co. K, 4th RI Vols. Sept. 30, 1861. Mustered out July 13, 1865.

Bliven, Benjamin C. Residence, Newport. 23. S. Carpenter. Enlisted in Co. K, 4th RI Vols. Aug. 13, 1862. Absent sick at Portsmouth Grove, RI, Mar. 1865, and so borne until mustered out June 16, 1865. Died Feb. 29, 1892. Interred at Oak Grove Cemetery, Fall River, MA.

Brown, William. Residence, Providence. 19. S. Farmer. Enlisted in Co. K, 4th RI Vols. Sept. 25, 1861. Mustered out July 13, 1865.

Burdick, Charles E. Residence, Newport. 19. S. Plumber. Enlisted in Co. B, 4th RI Vols. Aug. 13, 1862. Mustered out June 9, 1865.

Carey, Edward. Residence, Providence. 42. M. Laborer. Enlisted in Co. B, 4th RI Vols. June 23, 1862. Mustered out June 9, 1865.

Chase, Artemas B. Residence, New Bedford, MA. Enlisted in Co. B, 4th RI Vols. Sept. 13, 1861. Mustered out July 13, 1865. Died of disease contracted in the service at New Bedford, Massachusetts, Dec. 15, 1865.

Chase, Charles A. Residence, Woonsocket. 19. S. Clerk Enlisted in Co. E, 4th RI Vols. Sept. 17, 1862. Mustered out June 9, 1865. Died Aug. 6, 1911. Interred at Elm Grove Cemetery, North Kingstown, RI.

Cherry, Moses. Transferred from Co. A, June 9, 1865. Mustered out July 13, 1865. Interred at St. Patrick's Cemetery, East Greenwich, RI.

Doherty, Bernard. Transferred from Co. C, June 9, 1865. Mustered out July 13, 1865.

Donnelly, James. Residence, Woonsocket. 22. S. Operative. Enlisted in Co. E, 4th RI Vols. Sept. 19, 1861. Promoted to corporal July 1, 1865.

Driscoll, John A. Residence, Taunton, MA. 24. S. Shoemaker. Enlisted in Co. K, 4th RI Vols. Aug. 15, 1862. Mustered out July 13, 1865. Died 1914. Interred at Hope Cemetery, Kennebunk, ME.

Dunn, James L. Residence, Woonsocket. 18. S. Laborer. Enlisted in Co. E, 4th RI Vols. Aug. 2, 1862. Mustered out June 9, 1865.

Dunn, William. Residence, Woonsocket. 21. S. Laborer. Enlisted in Co. E, 4th RI Vols. Aug. 2, 1862. Mustered out June 21, 1865.

Easterbrooks, William H. Residence, Bristol. 19. S. Farmer. Enlisted in Co. K, 4th RI Vols. Sept. 23, 1861. Mustered out July 13, 1865. Died 1925. Interred at North Burial Ground, Bristol, RI.

Eckles, Edward. Residence, Newport. 38. M. Laborer. Enlisted in Co. K, 4th RI Vols. Sept. 11, 1861. Mustered out July 13, 1865. Interred at Hampton National Cemetery, Section FII, Grave 6067.

Ennis, Lawrence. Residence, Woonsocket. 22. M. Operative. Enlisted in Co. E, 4th RI Vols. July 29, 1862. Mustered out June 9, 1865. Died July 20, 1890. Interred at St. Charles Cemetery, Blackstone, MA.

Fiske, Eugene. Residence, Providence. 24. S. Laborer. Enlisted in Co. B, 4th RI Vols. April 19, 1864. Mustered in April 27, 1864. Mustered out July 13, 1865.

Flood, John. Residence, Smithfield. 18. S. Carder. Enlisted in Co. E, 4th RI Vols. Nov. 11, 1862. Mustered out July 13, 1865.

Griffith, Joseph H. Residence, Providence. 18. S. Farmer. Enlisted in Co. K, 4th RI Vols. Oct. 7, 1861. Mustered out July 13, 1865. Died Feb. 5, 1888. Interred at North Burial Ground, Providence, RI.

Hamilton, Robert. Residence, Woonsocket. 26. S. Mariner. Enlisted in Co. E, 4th RI Vols. Aug. 12, 1862. Mustered out July 13, 1865.

Hayes, John. Residence, Providence. 33. S. Laborer. Enlisted in Co. K, 4th RI Vols. Sept. 17, 1861. Mustered out July 13, 1865. Interred at St. Mary's Cemetery, West Warwick, RI.

Hennessey, William. Residence, Woonsocket. 25. S. Operative. Enlisted in Co. E, 4th RI Vols. Sept. 26, 1861. Mustered out July 13, 1865. Interred at St. Francis Cemetery, Pawtucket, RI.

Howard, George. Residence, Rochester, MA. 18. S. Farmer. Enlisted in Co. K, 4th RI Vols. Sept. 25, 1861. Mustered out July 13, 1865.

Howard, Michael. Residence, Newport. 34. S. Laborer. Enlisted in Co. G, 4th RI Vols. Sept. 16, 1861. Mustered out July 13, 1865.

Kelley, Joseph. Residence, Smithfield. 19. S. Operative. Enlisted in Co. E, 4th RI Vols. Sept. 10, 1861. Mustered out June 9, 1865.

Kelley, Thomas. Residence, Woonsocket. 19. S. Shoemaker Enlisted in Co. E, 4th RI Vols. Sept. 10, 1861. Mustered out July 13, 1865.

Kinney, James. Residence, Hopkinton. 31. M. Farmer. Enlisted in Co. B, 4th RI Vols. Sept. 23, 1861. Mustered out July 13, 1865. Interred at Wood River Cemetery, Richmond, RI.

Langland, Isaac. Transferred from Co. C, June 9, 1865. Mustered out July 13, 1865.

Lord, Charles. Residence, York, ME. 28. M. Machinist. Enlisted in Co. K, 4th RI Vols. Oct. 1, 1861. Mustered out July 13, 1865.

Lynch, Edwin. Residence, Cranston. 23. S. Farmer. Enlisted in Co. E, 4th RI Vols. Nov. 12, 1862. Mustered out July 13, 1865.

McNally, James. Residence, Woonsocket. 28. S. Hostler. Enlisted in Co. E, 4th RI Vols. Sept. 10, 1861. Mustered out July 13, 1865. Interred at Togus National Cemetery. Grave 2050.

Masterson, Patrick. Residence, Newport. 21. S. Laborer. Enlisted in Co. K, 4th RI Vols. Sept. 11, 1861. Mustered out July 13, 1865.

Murray, John J. Residence, Woonsocket. 26. S. Laborer. Enlisted in Co. E, 4th RI Vols. Sept. 10, 1861. Promoted to corporal, July 1, 1865.

Nickerson, Horace M. Residence, Woonsocket. 22. S. Teamster. Enlisted in Co. E, 4th RI Vols. Sept. 10, 1861. Mustered out July 13, 1865.

Ormes, Michael. Residence, Woonsocket. 19. S. Teamster. Enlisted in Co. E, 4th RI Vols. Sept. 10, 1861. Mustered out July 13, 1865.

Owens, Thomas T. Transferred from Co. C, June 9, 1865. Mustered out July 13, 1865. Died 1918. Interred at Manchester Cemetery, Coventry, RI.

Pickering, Henry W. Residence, Cumberland. 21. S. Farmer. Enlisted in Co. E, 4th RI Vols. Sept. 10, 1861. Mustered out July 13, 1865. Died Nov. 20, 1928. Interred at Wilcox Cemetery, Bellingham, MA.

Pierce, William F. Residence, Woonsocket. 43. M. Laborer. Enlisted in Co. E, 4th RI Vols. Sept. 10, 1862. Wagon master at corps headquarters Nov. 1864 until June 1865. Mustered out June 9, 1865. Died Nov. 30, 1874. Interred at Union Cemetery, North Smithfield, RI.

Potter, Henry M. Residence, Smithfield. 19. S. Laborer. Enlisted in Co. E, 4th RI Vols. Sept. 10, 1861. Mustered out July 13, 1865. Died Aug. 29, 1916. Interred at Moshassuck Cemetery, Central Falls, RI.

Preston, George W. Residence, Coventry. 30. S. Laborer. Enlisted in Co. B, 4th RI Vols. Sept. 5, 1861. Wounded in action at Petersburg, VA, April 2, 1865. Sent to hospital and borne as absent sick until June 21, 1865 when he reported from hospital. Mustered out July 29, 1865. Died Oct. 22, 1867. Interred at Moshassuck Cemetery, Central Falls, RI.

Regan, John. Residence, Woonsocket. 24. M. Weaver. Enlisted in Co. E, 4th RI Vols. Oct. 30, 1862. Mustered out July 13, 1865.

Reardon, Patrick. Residence, Providence. 22. S. Laborer. Enlisted in Co. G, 4th RI Vols. Sept. 11, 1861. Wounded in action at Petersburg, VA, April 2, 1865. Mustered out July 13, 1865.

Riley, James E. Residence, Newport. 23. S. Farmer. Enlisted in Co. G, 4th RI Vols. Sept. 11, 1861. Mustered out July 13, 1865. Died May 8, 1911. Interred at Waverly Cemetery, Waverly, KS.

Schofield, Joseph. Residence, Whitinsville, MA. 41. M. Machinist. Enlisted in Co. E, 4th RI Vols. Aug. 13, 1862. Mustered out June 9, 1865. Died April 12, 1879. Interred at Pine Grove Cemetery, Northbridge, MA.

Shay, Jeremiah. Residence, Newport. 35. M. Laborer. Enlisted in Co. K, 4th RI Vols. Sept. 11, 1861. Mustered out July 13, 1865.

Sheldon, David A. Residence, Johnston. 18. S. Farmer. Enlisted in Co. B, 4th RI Vols. Sept. 5, 1861. Mustered out July 13, 1865. Died 1915. Interred at Plainland Cemetery, Coventry, RI.

Sheldon, Lowell. Residence, Burrillville. 32. S. Laborer. Enlisted in Co. E, 4th RI Vols. Sept. 10, 1861. Mustered out July 13, 1865. Died May 20, 1902. Interred at Evergreen Cemetery, Douglas, MA.

Sherman, David. Transferred from Co. C, June 9, 1865. Mustered out July 13, 1865. Interred at Mineral Spring Cemetery, Pawtucket, RI.

Shippee, William J. Residence, Bristol. 29. S. Farmer. Enlisted in Co. K, 4th RI Vols. Sept. 24, 1861. Mustered out July 13, 1865. Died 1877. Interred at North Burial Ground, Bristol, RI.

Slocum, Charles F. Residence, Cranston. 21. S. Clerk. Enlisted in Co. K, 4th RI Vols. Sept. 23, 1861. Mustered out July 13, 1865. Died Sept. 23, 1879. Interred at Lakewood Burial Ground, Warwick, RI.

Smith, David C. Residence, Middletown. 21. S. Farmer. Enlisted in Co. K, 4th RI Vols. July 31, 1862. Mustered in Aug. 30, 1862. Mustered out July 13, 1865. Died 1925. Interred at Middletown Cemetery, Middletown, RI.

Smith, Edwin M. Residence, Smithfield. 18. S. Farmer. Enlisted in Co. E, 4th RI Vols. Sept. 16, 1861. Mustered out July 13, 1865. Interred at Greenville Cemetery, Smithfield, RI.

Steadman, William H. Transferred from Co. C, June 9, 1865. Mustered out July 13, 1865. Died 1927. Interred at Riverbend Cemetery, Westerly, RI.

Sullivan, Daniel. Residence, Newport. 21. S. Gardiner. Enlisted in Co. G, 4th RI Vols. Sept. 11, 1861. Mustered out July 13, 1865.

Sullivan, Martin. Residence, Smithfield. 21. S. Laborer. Enlisted in Co. E, 4th RI Vols. Sept. 19, 1861. Mustered out July 13, 1865.

Thayer, William T. Residence, Smithfield. 45. S. Farmer. Enlisted in Co. B, 4th RI Vols. Nov. 3, 1862. Mustered out July 13, 1865.

Thornley, Richard. Transferred from Co. C, June 9, 1865. Mustered out July 13, 1865. Interred at First Cemetery, East Greenwich, RI.

Waterman, George G. Residence, Johnston. 18. S. Laborer. Enlisted in Co. B, 4th RI Vols. July 25, 1864. Mustered out July 13, 1865. Died July 26, 1936. Interred at Lakeside-Carpenter Cemetery, East Providence, RI.

Waterman, William A. Residence, Providence. 26. M. Laborer. Enlisted in Co. B, 4th RI Vols. Sept. 5, 1861. Mustered out July 13, 1865. Died Feb. 25, 1900. Interred at St. Ann Cemetery, Cranston, RI.

Welch, James. Residence, Smithfield. 18. S. Spinner. Enlisted in Co. E, 4th RI Vols. Nov. 12, 1862. Mustered out July 13, 1865. Died Dec. 17, 1916. Interred at St. Paul's Cemetery, Blackstone, MA.

Weldon, George W. Transferred from Co. C, June 9, 1865. Mustered out July 13, 1865.

Weldon, Henry. Residence, Providence. 27. S. Mechanic. Enlisted in Co. E, 4th RI Vols. Sept. 27, 1861. Mustered out July 13, 1865. Interred at Grace Church Cemetery, Providence, RI.

COMPANY D (New Organization)

Captains

Brown, Edward P. Residence, Rehoboth, MA. Enlisted in 4th RI Vols. Sept. 1, 1861. Brevet major for heroism at storming of Petersburg, April 2, 1865. Resigned June 5, 1865. Died July 26, 1909. Interred at Woodlawn Cemetery, Bronx, NY.

Moore, Winthrop A. Promoted from first lieutenant Co. D, June 15, 1865. Mustered out July 13, 1865. Died Sept. 8, 1913. Interred at First Cemetery, East Greenwich, RI.

First Lieutenants

Moore, Winthrop A. Transferred from Co. K, Feb. 1, 1865. Promoted to captain Co. D, June 15, 1865.

Costello, George B. Promoted from second lieutenant Co. B, June 15, 1865. Mustered out July 13, 1865. Died of tuberculosis contracted in the service at Providence, RI, July 21, 1868. Interred at North Burial Ground, Providence, RI.

Second Lieutenant

Weeden, Merchant H. Promoted from sergeant Co. D, June 17, 1865. Mustered out July 13, 1865.

First Sergeant

Herbert, John C. Residence, Providence. 26. M. Carpenter. Enlisted in Co. F, 4th RI Vols. Sept. 14, 1861. Mustered out July 13, 1865.

Sergeants

Chase, William T. Residence, Providence. 23. M. Farmer. Enlisted in Co. F, 4th RI Vols. Sept. 14, 1861. Mustered out July 13, 1865. Died April 18, 1904. Interred at Old North Cemetery, Newport, RI.

Coggeshall, John S. Promoted from corporal July 1, 1865. Mustered out July 13, 1865. Died July 1, 1914. Interred at Oak Grove Cemetery, Fall River, MA.

Croning, Dennis. Residence, Providence. 21. S. Machinist. Enlisted in Co. C, 4th RI Vols. Sept. 9, 1861. Wounded in action at Petersburg, VA, Mar. 25, 1865. Mustered out July 13, 1865.

Harrington, William E. Promoted from corporal. Mustered out July 13, 1865.

McGahey, James Promoted from corporal. Mustered out July 13, 1865.

Weeden, Merchant H. Residence, Providence. 28. S. Carpenter. Enlisted in Co. F, 4th RI Vols. Sept. 14, 1861. Promoted to second lieutenant Co. D, June 17, 1865.

Corporals

Briggs, Edward C. Promoted from private. Mustered out July 13, 1865.

Brown, John H. Promoted from private. Mustered out July 13, 1865.

Coggeshall, John S. Residence, Tiverton. 18. S. Farmer. Enlisted in Co. F, 4th RI Vols. Sept. 17, 1861. Wounded in action at Petersburg, VA, Mar. 29, 1865. Promoted to sergeant July 1, 1865.

Douglas, George L. Residence, Scituate. 21. S. Operative. Enlisted in Co. C, 4th RI Vols. Sept. 9, 1861. Mustered out July 13, 1865. Died 1921. Interred at Hopkins Mill Cemetery, Foster, RI.

Grinell, Edson. Promoted from private. Mustered out July 13, 1865.

Harrington, William E. Residence, Providence. 18. S. Butcher. Enlisted in Co. C, 4th RI Vols. Sept. 9, 1861. Promoted to sergeant June 10, 1865.

Harkness, Charles. Promoted from private. Mustered out July 13, 1865.

Holley, William. Promoted from private. Mustered out July 13, 1865.

McGahey, James Residence, Providence. 18. S. Jeweler. Enlisted in Co. F, 4th RI Vols. Sept. 12, 1861. Promoted to sergeant June 10, 1865.

Malarkey, Charles. Promoted from private. Mustered out July 13, 1865.

Pitts, George. Residence, Providence. 22. M. Carpenter. Enlisted in Co. C, 4th RI Vols. Aug. 15, 1862. Mustered out June 9, 1865. Died Mar. 4, 1906. Interred at Locust Grove Cemetery, Providence, RI.

Rogers, Isaac H. Residence, Providence. 18. S. Baker. Enlisted in Co. F, 4th RI Vols. Sept. 17, 1861. Mustered out July 13, 1865. Died 1904. Interred at North Burial Ground, Providence, RI.

Shepley, John. Residence, Newport. 19. S. Laborer. Enlisted in Co. I, 4th RI Vols. Aug. 13, 1862. Mustered out June 9, 1865.

Wood, Henry W. Promoted from private. Mustered out July 13, 1865. Died Sept. 2, 1896. Interred at Pocasset Cemetery, Cranston, RI.

Musician

Kent, Benjamin D. Residence, Burrillville. 33. M. Painter. Enlisted in Co. A, 4th RI Vols. Sept. 28, 1861. Absent sick at Washington Mar. 1865. Mustered out July 13, 1865.

Privates

Armstrong, John. Residence, Providence. 19. S. Laborer. Enlisted in Co. C, 4th RI Vols. Sept. 16, 1864. Mustered out July 13, 1865. Died Sept. 10, 1897. Interred at St. Mary's Cemetery, New Bedford, MA.

Baacke, George E. Transferred from Co. H, June 9, 1865. Mustered out July 13, 1865.

Baggen, Bernard. Transferred from Co. H, June 9, 1865. Mustered out July 13, 1865.

Bagley, John. Residence, Cranston. 23. M. Operative. Enlisted in Co. F, 4th RI Vols. Sept. 17, 1861. Mustered out July 13, 1865. Died Oct. 12, 1915. Interred at St. Mary's Cemetery, West Warwick, RI.

Bentley, Allen W. Residence, Scituate. 23. M. Operative. Enlisted in Co. F, 4th RI Vols. Sept. 17, 1861. Mustered out July 13, 1865.

Beaumont, John. Residence, Cumberland. 25. S. Mule spinner. Enlisted in Co. F, 4th RI Vols. Oct. 30, 1862. Mustered out July 13, 1865. Interred at Oak Grove Cemetery, Pawtucket, RI.

Booth, William J. Transferred from Co. F, June 9, 1865. Mustered out July 13, 1865. Died May 10, 1919. Interred at North Burial Ground, Providence, RI.

Brennan, Thomas. Transferred from Co. E, June 9, 1865. Mustered out July 13, 1865.

Briggs, Edward C. Residence, Providence. 35. M. Laborer. Enlisted in Co. C, 4th RI Vols. Sept. 9, 1861. Promoted to corporal July 1, 1865.

Briggs, Nathan. Residence, Coventry. 36. M. Farmer. Enlisted in Co. C, 4th RI Vols. Aug. 5, 1862. Mustered out June 9, 1865. Died Sept. 17, 1894. Interred at Knotty Oak Cemetery, Coventry, RI.

Brown, John H. Residence, Providence. 19. S. Stone cutter. Enlisted in Co. F, 4th RI Vols. Sept. 16, 1861. Promoted to corporal June 10, 1865.

Carroll, Thomas. Residence, Tiverton. 29. M. Laborer. Enlisted in Co. F, 4th RI Vols. Sept. 17, 1861. Mustered out July 13, 1865.

Clavin, Michael. Transferred from Co. F, June 9, 1865. Mustered out July 13, 1865. Died Dec. 29, 1867. Interred at St. Patrick's Cemetery, Burrillville, RI.

Coen, James. Residence, Providence. 33. M. Teamster. Enlisted in Co. F, 4th RI Vols. Sept. 14, 1861. Mustered out July 13, 1865. Died May 3, 1883. Interred at Togus National Cemetery. Grave 361.

Coffey, Peter. Residence, Providence. 18. S. Laborer. Enlisted in Co. I, 4th RI Vols. Mar. 11, 1864. Mustered out July 13, 1865.

Covill, George W. Transferred from Co. H, June 9, 1865. Mustered out July 13, 1865. Died Dec. 12, 1902. Interred at Pine Grove Cemetery, Coventry, RI.

Connelly, Simeon. Transferred from Co. E, June 9, 1865. Mustered out July 13, 1865.

Corey, John A. Residence, Richmond. 25. S. Laborer. Enlisted in Co. F, 4th RI Vols. Sept. 12, 1861. Mustered out July 13, 1865. Died Feb. 4, 1917. Interred at Pine Grove Cemetery, Hopkinton, RI.

Dorman, William. Residence, Providence. 34. M. Laborer. Enlisted in Co. F, 4th RI Vols. Sept. 17, 1861. Mustered out July 13, 1865. Died Jan. 11, 1883. Interred at Saints Peter & Paul Cemetery, Coventry, RI.

Dow, Byron E. Residence, Providence. 21. S. Machinist. Enlisted in Co. C, 4th RI Vols. Sept. 9, 1861. Mustered out July 13, 1865. Died June 16, 1869 "of disease contracted in the U.S. service during the Great Rebellion." Interred at Locust Grove Cemetery, Providence, RI.

Duffy, Michael. Residence, Warwick. 24. S. Laborer. Enlisted in Co. F, 4th RI Vols. Sept. 17, 1861. Mustered out July 13, 1865. Died Dec. 27, 1867. Interred at St. Mary's Cemetery, West Warwick, RI.

Fieldson, Joseph. Residence, Providence. 21. S. Operative. Enlisted in Co. F, 4th RI Vols. Sept. 14, 1861. Mustered out July 13, 1865.

Fieldson, Joshua. Residence, Woonsocket. 44. M. Laborer. Enlisted in Co. F, 4th RI Vols. Sept. 14, 1861. Mustered out July 13, 1865.

Foster, James A. Transferred from Co. H, June 9, 1865. Mustered out July 13, 1865. Died May 15, 1919. Interred at Swan Point Cemetery, Providence, RI.

Gill, John. Transferred from Co. E, June 9, 1865. Mustered out July 13, 1865. Died 1902. Interred at Moshassuck Cemetery, Central Falls, RI.

Gladding, Nathaniel W. Transferred from Co. H, June 9, 1865. Mustered out July 13, 1865.

Gladding, Oliver H. P. Residence, Newport. 19. S. Painter. Enlisted in Co. F, 4th RI Vols. Aug. 15, 1862. Mustered out June 9, 1865.

Goddard, Joseph Jr. Transferred from Co. D, June 9, 1865. Mustered out July 13, 1865.

Gorton, Elisha. Transferred from Co. D, June 9, 1865. Mustered out July 13, 1865.

Greene, Daniel H. Residence, Providence. 19. S. Student. Enlisted in Co. I, 4th RI Vols. Sept. 19, 1861. Mustered out July 21, 1865. Died Mar. 11, 1911. Interred at North Burial Ground, Providence, RI.

Grinnell, Edson. Residence, Tiverton. 23. S. Farmer. Enlisted in Co. F, 4th RI Vols. Sept. 23, 1861. Promoted to corporal June 10, 1865.

Hanley, Thomas. Residence, Providence. 25. S. Operative. Enlisted in Co. C, 4th RI Vols. Dec. 2, 1862. Mustered out July 13, 1865. Interred at St. Francis Cemetery, Pawtucket, RI.

Harkness, Charles. Residence, Roxbury, MA. 25. S. Brass finisher. Enlisted in Co. F, 4th RI Vols. Sept. 14, 1861. Promoted to corporal June 10, 1865.

Hill, Albert H. Residence, Scituate. 21. S. Operative. Enlisted in Co. C, 4th RI Vols. July 1, 1862. Mustered out June 9, 1865. Died Sept. 24, 1901. Interred at Union Cemetery, Moosup, CT.

Hill, Baxter M. Transferred from Co. H, June 9, 1865. Mustered out July 13, 1865. Died April 24, 1880. Interred at North Burial Ground, Providence, RI.

Hill, Jerry. Residence, Scituate. 24. M. Operative. Enlisted in Co. C, 4th RI Vols. July 1, 1862. Mustered out July 13, 1865.

Hill, Joseph N. Residence, Providence. 21. S. Farmer. Enlisted in Co. C, 4th RI Vols. July 1, 1862. Mustered out June 9, 1865. Died 1894. Interred at Riverside Cemetery, Hiram Rapids, OH.

Holley, William. Residence, Providence. 18. S. Laborer. Enlisted in Co. F, 4th RI Vols. Sept. 17, 1861. Promoted to corporal, June 10, 1865.

Holland, John. Residence, Providence. 23. S. Blacksmith. Enlisted in Co. F, 4th RI Vols. Sept. 2, 1862. Mustered out July 13, 1865. Interred at St. Francis Cemetery, Pawtucket, RI.

Holloway, Elisha. Transferred from Co. H, June 9, 1865. Mustered out July 13, 1865. Died Mar. 10, 1897. Interred at Clarke-Webster Lot, Charlestown Cemetery 46, Charlestown, RI.

Hunt, Daniel D. Residence, North Kingstown. 19. S. Farmer. Enlisted in Co. I, 4th RI Vols. Mar. 11, 1864. Mustered out at Washington, DC, July 5, 1865. Died Dec. 7, 1898. Interred at Rufus-Hunt Cemetery, North Kingstown Cemetery 10, North Kingstown, RI.

Hunt, Leonard A. Residence, North Kingstown. 18. S. Farmer. Enlisted in Co. I, 4th RI Vols. April 11, 1864. Mustered out July 13, 1865. Interred at Mt. Hope Cemetery, Swansea, MA.

Kettle, Charles A. Residence, Coventry. 33. S. Spinner. Enlisted in Co. C, 4th RI Vols. Aug. 4, 1862. Died of dysentery in hospital at Camp Parole, Annapolis, MD, Mar. 19, 1865. Interred at Annapolis National Cemetery. Section C, Grave 1068.

Lake, Israel F. Jr. Residence, Newport. 28. S. Laborer. Enlisted Mar. 8, 1865. Mustered out July 13, 1865. Died Mar. 16, 1903. Interred at Island Cemetery, Newport, RI.

Lee, Frank. Transferred from Co. H, June 9, 1865. Mustered out July 13, 1865.

Longstreet, Frank B. Residence, Providence. 17. S. Clerk. Enlisted in Co. I, 4th RI Vols. May 19, 1862. Mustered out May 13, 1865.

Lynch, Daniel. Residence, Providence. 30. M. Laborer. Enlisted in Co. C, 4th RI Vols. Sept. 9, 1861. Wounded in action at Petersburg, VA, April 2, 1865. Mustered out July 13, 1865. Died Jan. 11, 1873. Interred at St. Patrick's Cemetery, Providence, RI.

McCann, James. Residence, Smithfield. 42. M. Operative. Enlisted in Co. C, 4th RI Vols. Nov. 1, 1862. Mustered out July 13, 1865.

McGill, John. Residence, Providence. 25. M. Laborer. Enlisted in Co. I, 4th RI Vols. Mar. 9, 1864. Discharged for disability June 10, 1865.

McHugh, Peter. Residence, Providence. 18. S. Sailor. Enlisted in Co. C, 4th RI Vols. Mar. 11, 1864. Mustered out July 13, 1865. Interred at St. Francis Cemetery, Pawtucket, RI.

McLaughlin, Charles. Residence, Providence. 18. S. Laborer. Enlisted April 10, 1865. Mustered out July 13, 1865.

McQuernan, Terrence. Residence, Providence. 18. S. Laborer. Enlisted Mar. 9, 1865. Mustered out July 13, 1865.

McShane, Patrick. Residence, Boston, MA. 35. M. Laborer. Enlisted in Co. F, 4th RI Vols. Sept. 17, 1861. Wounded in action at Petersburg, VA, April 2, 1865. Mustered out July 13, 1865. Died 1886. Interred at St. Mary's Cemetery, West Warwick, RI.

Malarkey, Charles. Residence, Providence. 21. S. Laborer Enlisted in Co. F, 4th RI Vols. Sept. 17, 1861. Promoted to corporal.

Moody, John. Residence, Providence. 31. M. Farmer. Enlisted in Co. C, 4th RI Vols. Sept. 9, 1861. Mustered out July 13, 1865.

Moon, Oliver. Residence, Coventry. 34. M. Farmer. Enlisted in Co. C, 4th RI Vols. Aug. 2, 1862. Mustered out June 9, 1865. Interred at Knotty Oak Cemetery, Coventry, RI.

Morrissey, John. Transferred from Co. H, June 9, 1865. Mustered out July 13, 1865

O'Dell, George W. Residence, New York, NY. 21. S. Teamster. Enlisted in Co. C, 4th RI Vols. Sept. 9, 1861. Transferred to Veteran Reserve Corps.

Olsen, Henry. Residence, New York, NY. 20. S. Farmer. Enlisted in Co. C, 4th RI Vols. Jan. 22, 1863. Mustered out July 13, 1865.

Ormsbee, William W. Residence, Richmond. 19. S. Laborer. Enlisted in Co. C, 4th RI Vols. Sept. 9, 1861. Mustered out July 13, 1865.

Pitts, Joseph. Residence, Providence. 18. S. Jeweler. Enlisted in Co. C, 4th RI Vols. Aug. 15, 1862. Mustered out June 9, 1865. Died Oct. 28, 1888. Interred at Pelham Cemetery, New Harmony, IN.

Prestwick, Thomas. Residence, Providence. 17. S. Jeweler. Enlisted in Co. C, 4th RI Vols. Sept. 9, 1861. Mustered out July 13, 1865.

Quinn, John. Residence, Providence. 25. S. Laborer. Enlisted in Co. I, 4th RI Vols. Sept. 21, 1861. Mustered out July 13, 1865.

Riley, Peter. Residence, Providence. 18. S. Laborer. Enlisted in Co. I, 4th RI Vols. Aug. 7, 1862. Mustered out June 9, 1865.

Searle, Franklin V. Residence, Scituate. 18. S. Laborer. Enlisted in Co. F, 4th RI Vols. Sept. 17, 1861. Mustered out July 13, 1865. Died 1902. Interred at Barden Cemetery, Scituate Cemetery 39, Scituate, RI.

Smith, James. Transferred from Co. H, June 9, 1865. Mustered out July 13, 1865.

Sullivan, Michael. Residence, Providence. 18. S. Bootmaker. Enlisted in Co. C, 4th RI Vols. Nov. 8, 1862. Mustered out July 13, 1865.

Sunderland, William N. Residence, Warwick. 21. M. Baker. Enlisted in Co. I, 4th RI Vols. Sept. 21, 1861. Mustered out July 13, 1865. Died May 2, 1896. Interred at Togus National Cemetery. Grave 1269.

Thomas, Elisha. Residence, Cranston. 44. M. Farmer. Enlisted in Co. I, 4th RI Vols. Aug. 28, 1862. Mustered out June 9, 1865.

Tiernan, John B. Residence, Boston, MA. 19. S. Bootmaker. Enlisted in Co. C, 4th RI Vols. July 27, 1864. Mustered out July 13, 1865. Died Aug. 4, 1918. Interred at Granger Fairview Cemetery, Granger, OH.

Tillinghast, Charles E. Residence, Warwick. 18. S. Laborer. Enlisted in Co. F, 4th RI Vols. Sept. 16, 1861. Mustered out July 13, 1865. Died Feb. 5, 1908. Interred at Topeka Cemetery, Topeka, KS.

Trainor, Michael. Transferred from Co. D, June 9, 1865. Died of dysentery at Second Division, Ninth Army Corps Hospital, Washington, DC, July 7, 1865. Interred at Alexandria National Cemetery. Grave 3229.

Travers, Thomas. Residence, Providence. 24. M. Farmer. Enlisted in Co. C, 4th RI Vols. Mar. 22, 1864. Mustered out July 13, 1865.

Viall, William S. Residence, Rehoboth, MA. 22. S. Farmer. Enlisted in Co. C, 4th RI Vols. Sept. 9, 1861. Mustered out July 13, 1865. Died Sept. 12, 1890. Interred at Harrington Cemetery, Cuyahoga Falls, OH.

Whitman, Reuben A. Residence, Warwick. Enlisted in Co. F, 4th RI Vols. Sept. 14, 1861. Died of dysentery in Second Division, Ninth Army Corps Hospital, City Point, VA, Mar. 20, 1865. Interred at City Point National Cemetery. Section C, Grave 2190. Cenotaph in Greenwood Cemetery, Coventry, RI.

Whitman, Thomas R. Transferred from Co. F, June 9, 1865. Mustered out July 13, 1865. Died Nov. 6, 1881. Interred at Greenwood Cemetery, Coventry, RI.

Wilcox, Elijah R. Residence, Tiverton. 29. M. Carpenter. Enlisted in Co. F, 4th RI Vols. Aug. 15, 1862. Mustered out June 9, 1865. Died 1912. Interred at Wilcox Cemetery, Bellingham, MA.

Wilson, Joseph. Residence, Providence. 34. M. Enlisted in Co. C, 4th RI Vols. Oct. 3, 1863. Wounded in action, shot in head, at Petersburg, VA, April 2, 1865. Mustered out July 13, 1865. Died Sept. 23, 1896. Interred at North Burial Ground, Providence, RI.

Wood, Caleb G. Residence, Coventry. 22. M. Laborer. Enlisted in Co. C, 4th RI Vols. Aug. 4, 1862. Mustered out June 9, 1865. Died

Nov. 30, 1911. Interred at Large Maple Root Cemetery, Coventry, RI.

Wood, Henry W. Residence, Providence. 17. S. Student. Enlisted in Co. F, 4th RI Vols. Sept. 16, 1861. Promoted to corporal June 10, 1865. Died Sept. 2, 1896. Interred at Pocasset Cemetery, Cranston, RI.

COMPANY G (New Organization)

Captain

Bowen, Caleb T. Residence, North Kingstown. 28. M. Merchant. Enlisted in Co. H, 4th RI Vols. Sept. 13, 1861. Mustered out July 13, 1865. Died Dec. 9, 1906. Interred at Lone Fir Pioneer Cemetery, Portland, OR.

First Lieutenant

Remington, Daniel S. Transferred from Co. C, Feb. 1, 1865. Promoted to captain Co. D, June 15, 1865.

First Sergeant

Groff, Charles E. Promoted from sergeant. Commissioned second lieutenant in June 1865, but never mustered as such. Mustered out July 13, 1865. Died 1906. Interred at North Burial Ground, Newport, RI.

Sergeants

Allen, Russell W. Promoted from corporal. Mustered out July 13, 1865. Died 1908. Interred at Mineral Spring Cemetery, Pawtucket, RI.

Earley, Patrick. Promoted from corporal. Mustered out July 13, 1865.

Groff, Charles E. Residence, Newport. 21. S. Carpenter. Enlisted in Co. G, 4th RI Vols. Sept. 11, 1861. Promoted to first sergeant Mar. 1, 1865.

Nottage, William N. Residence, Providence. 28. M. Jeweler. Enlisted in Co. H, 4th RI Vols. Sept. 24, 1861. Mustered out July

13, 1865. Died April 9, 1916. Interred at St. Patrick's Cemetery, Providence, RI.

Corporals

Allen, Russell W. Residence, Pawtucket. 19. S. Jeweler. Enlisted in Co. F, 4th RI Vols. Sept. 17, 1861. Promoted to sergeant June 10, 1865.

Barber, John F. Promoted from private. Mustered out July 13, 1865.

Bicknell, Thomas. Promoted from private. Mustered out July 13, 1865.

Colwell, Marcus M. Residence, Providence. 19. S. Jeweler. Enlisted in Co. A, 4th RI Vols. Sept. 7, 1861. Mustered out Aug. 25, 1865.

Earley, Patrick. Residence, Providence. 18. S. Machinist. Enlisted in Co. F, 4th RI Vols. Sept. 13, 1861. Promoted to sergeant June 10, 1865.

Lockwood, George A. Promoted from private. Mustered out July 13, 1865.

Shaw, George C. Promoted from private. Mustered out July 13, 1865. Died 1917. Interred at Island Cemetery, Newport, RI.

Sutherland, Andrew. Residence, Newport. 23. M. Shoe maker. Enlisted in Co. G, 4th RI Vols. Aug. 12, 1862. Mustered out July 13, 1865. Died Sept. 10, 1902. Interred at Island Cemetery, Newport, RI.

Weeden, Daniel W. Residence, Jamestown. 27. M. Farmer. Enlisted in Co. G, 4th RI Vols. Sept. 11, 1861. Mustered out July 13, 1865. Died Mar. 4, 1873. Interred at Holy Cross Episcopal Cemetery, Middletown, RI.

Musician

Smith, Albert J. Residence, Newport. 18. S. Musician. Enlisted in Co. G, 4th RI Vols. Sept. 7, 1861. Mustered out July 13, 1865. Died Aug. 14, 1887. Interred at Common Burying Ground, Newport, RI.

Privates

Andrews, James H. Residence, Glocester. 18. S. Laborer. Enlisted in Co. D, 4th RI Vols. May 28, 1862. Mustered out July 13, 1865. Died Mar. 9, 1885. Interred at Place-Keach Lot, Glocester Cemetery 25, Glocester, RI.

Arnold, Gilbert H. Residence, Burrillville. 21. M. Laborer. Enlisted in Co. D, 4th RI Vols. Aug. 13, 1862. Mustered out June 9, 1865. Died Jan. 2, 1910. Interred at Union Cemetery, North Smithfield, RI.

Babcock, George Manton. Residence, Providence. Enlisted in Co. G, 4th RI Vols. May 15, 1862. Mustered out at Alexandria, VA, May 14, 1865. Interred at Swan Point Cemetery, Providence, RI.

Barber, John F. Residence, Providence. 20. S. Gilder. Enlisted in Co. A, 4th RI Vols. Sept. 5, 1861. Promoted to corporal June 10, 1865.

Barber, William. Residence, Newport. 27. M. Shoemaker. Enlisted in Co H, 4th RI Vols. Aug. 15, 1862. Mustered out June 9, 1865. Interred at Island Cemetery, Newport, RI.

Bassett, George E. Residence, Smithfield. 18. S. Laborer. Enlisted in Co. H, 4th RI Vols. Aug. 7, 1862. Mustered out June 9, 1865.

Bradley, Abraham. Residence, Burrillville. 22. S. Weaver. Enlisted in Co. D, 4th RI Vols. Aug. 3, 1861. Mustered out July 13, 1865. Died 1927. Interred at St. Patrick Cemetery, Hogansburg, NY.

Briggs, Nathan O. Residence, Glocester. 25. S. Shoemaker. Enlisted in Co. D, 4th RI Vols. Sept. 16, 1861. Mustered out July 13, 1865. Died of tuberculosis contracted in the service June 6,

1867 at Putnam, CT. Interred at Grove Street Cemetery, Putnam, CT

Bicknell, Thomas. Residence, Providence. 21. S. Brush Maker. Enlisted in Co. A, 4th RI Vols. Sept. 6, 1861. Promoted to corporal Feb. 15, 1865.

Brayman, John R. Residence, Providence. 20. S. Operative. Enlisted in Co. A, 4th RI Vols. Sept. 5, 1861. Mustered out July 13, 1865.

Cahoone, Gideon A. Residence, Cranston. 21. S. Farmer. Enlisted in Co. D, 4th RI Vols. Aug. 2, 1862. Mustered out June 9, 1865. Died July 28, 1912. Interred at Knotty Oak Cemetery, Coventry, RI.

Carline, Patrick. Residence, Taunton, MA. 22. S. Shoemaker. Enlisted in Co. A, 4th RI Vols. Sept 5, 1861. Mustered out July 13, 1865.

Casey, Patrick. Residence, North Bridgewater, MA. 23. S. Shoemaker. Enlisted in Co. A, 4th RI Vols. Sept. 7, 1861. Mustered out July 13, 1865.

Clemence, George B. Residence, Burrillville. 20. S. Laborer. Enlisted in Co. D, 4th RI Vols. Aug. 7, 1861. Mustered out July 13, 1865. Died 1926. Interred at Acotes Hill Cemetery, Glocester, RI.

Cook, Isaac B. Residence, Tiverton. 19. S. Farmer. Enlisted in Co. A, 4th RI Vols. Aug. 13, 1862. Mustered out June 9, 1865. Died April 12, 1905. Interred at David Rounds Lot, Tiverton Cemetery 16, Tiverton, RI.

Donnelly, Frank. Residence, Providence. 41. M. Laborer. Enlisted in Co. A, 4th RI Vols. Sept. 6, 1861. Mustered out July 13, 1865.

Doonan, Frank. Residence, North Kingstown. 24. S. Laborer. Enlisted in Co. H, 4th RI Vols. Sept. 23, 1861. Mustered out July 13, 1865.

Donovan, Richard. Residence, North Kingstown. 22. S. Laborer. Enlisted in Co. H, 4th RI Vols. Sept. 13, 1861. Wounded in action at Petersburg, VA, April 2, 1865. Discharged for disability at Portsmouth Grove, RI, June 29, 1865.

Earley, John. Residence, Providence. 40. M. Farmer. Enlisted in Co. A, 4th RI Vols. Aug. 21, 1864. Mustered out July 13, 1865. Interred at St. Mary's Cemetery, Pawtucket, RI.

Flanders, Orlando D. Residence, Chelsea, MA. 18. S. Operative. Enlisted in Co. A, 4th RI Vols. Sept. 18, 1861. Mustered out July 13, 1865.

Flemming, David J. Transferred from Co. K, June 9, 1865. Mustered out July 13, 1865. Died Nov. 18, 1907. Interred at Slatersville Cemetery, North Smithfield, RI.

Freeborn, John P. Residence, Newport. 19. S. Sailor. Enlisted in Co. G, 4th RI Vols. Aug. 16, 1862. Absent sick Mar. 1865. Mustered out at Washington, DC, July 5, 1865. Died 1917. Interred at Middletown Village Cemetery, Middletown, RI.

Gavitt, Reynolds H. C. Residence, Providence. 18. S. Laborer. Enlisted in Co. A, 4th RI Vols. Aug. 7, 1862. Mustered out July 13, 1865. Interred at Wood River Cemetery, Richmond, RI.

Gorman, Morris. Residence, Milford, MA. 21. S. Shoemaker. Enlisted in Co. A, 4th RI Vols. Sept. 5, 1861. Mustered out July 13, 1865.

Gordon, Henry W. Residence, Coventry. 18. S. Farmer. Enlisted in Co. G, 4th RI Vols. Aug. 7, 1862. Died of disease at Fort Schuyler Hospital, NY, Feb. 1, 1865.

Hawkins, Robert S. Residence, Scituate. 18. S. Farmer. Enlisted in Co. D, 4th RI Vols. Aug. 27, 1861. Mustered out July 13, 1865.

Higgins, James. Residence, Providence. 21. S. Clerk Enlisted in Co. H, 4th RI Vols. Aug. 14, 1862. Mustered out June 9, 1865.

Hodson, Robert. Residence, Warwick. 33. M. Painter. Enlisted in Co. A, 4th RI Vols. Sept. 7, 1861. Mustered out July 13, 1865.

Hoxie, John W. Residence, East Greenwich. 21. S. Farmer. Enlisted in Co. H, 4th RI Vols. Sept. 23, 1861. Mustered out July 13, 1865.

Hunt, William H. Residence, North Kingstown. 21. S. Farmer. Enlisted in Co. H, 4th RI Vols. Sept. 19, 1861. Mustered out July 13, 1865. Died July 8, 1916. Interred at Elm Grove Cemetery, North Kingstown, RI.

Kelly, Malachi. Residence, Newport. 18. S. Laborer. Enlisted in Co. G, 4th RI Vols. Aug. 8, 1862. Mustered out June 9, 1865.

Kelley, Owen. Transferred from Co. I, June 9, 1865. Mustered out July 13, 1865. Died Mar. 22, 1896. Interred at St. Mary's Cemetery, Bristol, RI.

Kimball, Josiah H. Residence, Burrillville. 20. S. Teamster Enlisted in Co. D, 4th RI Vols. Aug. 6, 1861. Teamster at corps headquarters from Mar. 1865 to July 1865. Mustered out July 13, 1865. Died Nov. 5, 1928. Interred at Mt. Vernon Cemetery, Mt. Vernon, Skagit County, Washington.

Lacey, James. Residence, Burrillville. 22. S. Laborer. Enlisted in Co. D, 4th RI Vols. Aug. 31, 1861. Mustered out July 13, 1865. Died Aug. 23, 1931. Interred at St. Patrick's Cemetery, Burrillville, RI.

Leary, John. Residence, Warwick. 37. M. Seaman. Enlisted in Co. D, 4th RI Vols. Sept. 15, 1861. Mustered out July 13, 1865.

Leonard, George A. Residence, Providence. 29. S. Blacksmith. Enlisted in Co. A, 4th RI Vols. Sept. 6, 1861. Mustered out July 13, 1865. Died Oct. 24, 1890. Interred at North Burial Ground, Providence, RI.

Locklin, Thomas Jr. Transferred from Co. I, June 9, 1865. Mustered out July 13, 1865.

Lockwood, George A. Residence, Glocester. 37. M. Laborer. Enlisted in Co. D, 4th RI Vols. Aug. 12, 1861. Promoted to corporal June 10, 1865.

McCarty, Jeremiah. Residence, Newport. 23. M. Harness maker. Enlisted in Co. G, 4th RI Vols. Sept. 11, 1861. Mustered out July 13, 1865.

Markham, James. Residence, Newport. 32. S. Moulder. Enlisted in Co. G, 4th RI Vols. Sept. 11, 1861. Mustered out July 13, 1865.

Mallett, Michael. Residence, Boston, MA. 26. S. Laborer. Enlisted in Co. D, 4th RI Vols. Sept. 18, 1862. Mustered out June 9, 1865. Died Feb. 20, 1913. Interred at Washington State Veterans Home, Retsil, Washington.

Manning, Patrick. Residence, Providence. 29. M. Blacksmith. Enlisted in Co. A, 4th RI Vols. Sept. 5, 1861. Acting as brigade blacksmith, June 1865. Mustered out July 13, 1865. Interred at St. Mary's Cemetery, Newport, RI.

Mason, William, Jr. Residence, Newport. 22. S. Seaman. Enlisted in Co. G, 4th RI Vols. Aug. 16, 1862. Mustered out June 9, 1865. Interred at Union Village Cemetery, North Smithfield, RI.

Mitchell, Silas. Residence, New Shoreham. 22. S. Laborer. Enlisted in Co. H, 4th RI Vols. Aug. 6, 1862. In ambulance corps from Aug. 1864, until June 1865. Mustered out June 9, 1865. Died Feb. 22, 1891. Interred at Knotty Oak Cemetery, Coventry, RI.

Murphy, Patrick. Residence, Newport. 19. S. Laborer. Enlisted in Co. G, 4th RI Vols. Sept. 4, 1861. Mustered out July 13, 1865. Died 1891. Interred at St. Mary's Cemetery, Newport, RI.

Myers, Abraham. Residence, Stockbridge, MA. 19. S. Weaver. Enlisted in Co. K, 4th RI Vols. Sept. 19, 1861. Mustered out July 13, 1865. Died 1915. Interred at North Burial Ground, Bristol, RI.

Northrup, John R. Residence, North Kingstown. 27. M. Laborer. Enlisted in Co. H, 4th RI Vols. Sept. 13, 1861. Mustered out July 13, 1865.

Potter, Philip J. Residence, Glocester. 18. S. Laborer. Enlisted in Co. D, 4th RI Vols. Sept. 2, 1861. Mustered out July 13, 1865.

Phillips, Andrew J. 19. S. Residence, North Kingstown. Enlisted in Co. G, 4th RI Vols. Aug. 7, 1862. Mustered out June 9, 1865. Died 1920. Interred at Greenwood Cemetery, Coventry, RI.

Quigley, Martin. Residence, Smithfield. 25. S. Laborer. Enlisted in Co. D, 4th RI Vols. Sept. 21, 1861. Mustered out July 13, 1865. Died Jan. 30, 1922. Interred at St. Francis Cemetery, Pawtucket, RI.

Rose, Daniel R. Residence, North Kingstown. 22. S. Farmer. Enlisted in Co. H, 4th RI Vols. Sept. 23, 1861. Mustered out July 13, 1865. Interred at Elm Grove Cemetery, North Kingstown, RI.

Rourke, Walter. Residence, Providence. 18. S. Laborer. Enlisted in Co. A, 4th RI Vols. Sept. 6, 1861. Mustered out July 13, 1865.

Salisbury, George M. Residence, Pawtucket. 32. Enlisted in Co. A, 4th RI Vols. Sept. 5, 1861. Mustered out July 13, 1865.

Shaw, George C. Residence, Newport. 16. M. Laborer. Enlisted in Co. G, 4th RI Vols. Sept. 11, 1861. Promoted to corporal June 10, 1865.

Smith, Darius. Residence, North Kingstown. 35. S. Mariner. Enlisted in Co. H, 4th RI Vols. Sept. 13, 1861. Mustered out July 13, 1865. Died June 28, 1889. Interred at Elm Grove Cemetery, North Kingstown, RI.

Smith, Thomas A. Residence, Providence. 15. S. Laborer. Enlisted in Co. H, 4th RI Vols. Sept. 14, 1861. Mustered out July 13, 1865. Died May 7, 1911. Interred at Elm Grove Cemetery, North Kingstown, RI.

Smith, William H. Residence, Burrillville. 28. M. Laborer. Enlisted in Co. D, 4th RI Vols. Sept. 10, 1861. Mustered out July 13, 1865. Interred at Elm Grove Cemetery, North Kingstown, RI.

Spooner, Lovell T. Residence, Newport. 22. S. Weaver. Enlisted in Co. G, 4th RI Vols. Aug. 7, 1862. In commissary department from Mar. 1865, until June 1865. Mustered out June 9, 1865.

Sullivan, Timothy. Residence, Newport. 18. S. Laborer. Enlisted in Co. D, 4th RI Vols. Aug.17, 1862. Mustered out June 9, 1865. Died Sept. 24, 1879. Interred at St. Mary's Cemetery, Newport, RI.

Sweet, Herbert N. Residence, Smithfield. 18. S. Laborer. Enlisted in Co. D, 4th RI Vols. Sep. 16, 1861. Mustered out July 13, 1865. Died April 5, 1904. Interred at Willard Asylum Cemetery, Willard, NY.

Tanner, Richard D. Transferred from Co. I, June 9, 1865. Mustered out July 13, 1865. Died 1921. Interred at Locust Grove Cemetery, Providence, RI.

Taylor, Edward E. Residence, Newport. 30. M. Fisherman. Enlisted in Co. H, 4th RI Vols. Aug. 14, 1862. Mustered out June 9, 1865. Died April 12, 1866. Interred at Common Burial Ground, Newport, RI.

Tompkins, Franklin P. Residence, Providence. 28. M. Moulder. Enlisted in Co. A, 4th RI Vols. Sept. 5, 1861. Mustered out July 13, 1865. Died Feb. 4, 1896. Interred at Riverside Cemetery, Pawtucket, RI.

Troutz, George. Residence, Providence. 30. S. Jeweler. Enlisted in Co. H, 4th RI Vols. Sept. 20, 1861. Mustered out July 13, 1865.

Wall, William. Residence, Providence. 21. S. Laborer. Enlisted in Co. D, 4th RI Vols. June 16, 1862. Mustered out July 13, 1865. Died July 21 1888. Interred at St. Patrick's Cemetery, Providence, RI.

Watson, Elisha R. Residence, Coventry. 19. S. Farmer. Enlisted in Co. D, 4th RI Vols. Aug. 5, 1862. Mustered out June 9, 1865. Died April 20, 1939. Interred at Knotty Oak Cemetery, Coventry, RI.

Williams, John W. Residence, Newport. 18. S. Laborer. Enlisted in Co. G, 4th RI Vols. Sept. 11, 1861. Mustered out July 13, 1865. Died of dysentery contracted in the service at North Kingstown, RI, Oct. 14, 1865. Interred at Elm Grove Cemetery, North Kingstown, RI.

Wood, Horace B. Residence, Coventry. 19. S. Farmer. Enlisted in Co. A, 4th RI Vols. Nov. 19, 1862. Mustered out July 13, 1865. Died Feb. 18, 1913. Interred at Nunica Cemetery, Nunica, MI.

Willis, Jeremiah. Residence, North Kingstown. 19. S. Laborer. Enlisted in Co. H, 4th RI Vols. Sept. 18, 1862. Mustered out July 13, 1865. Died Jan. 6, 1910. Interred at Elm Grove Cemetery, North Kingstown, RI.

Whipple, William D. Residence, Providence. 18. S. Carpenter. Enlisted in Co. H, 4th RI Vols. Aug. 15, 1862. Mustered out June 9, 1865. Interred at Benedict Whipple Lot, Scituate Cemetery 78, Scituate, RI.

Register of Enlistment by Town

Town	Original Members and Recruits	Transferred from 4th R.I.V.
Barrington	4	0
Bristol	35	2
Burrillville	32	10
Charlestown	6	0
Coventry	28	9
Cranston	22	5
Cumberland	32	5
East Greenwich	26	1
East Providence	3	0
Exeter	23	0
Foster	24	0
Glocester	29	4
Hopkinton	54	1
Jamestown	0	1
Johnston	22	2
Little Compton	3	0
Middletown	0	1
New Shoreham	0	1
Newport	38	26
North Kingstown	23	12
North Providence	36	0
Pawtucket	25	2
Portsmouth	10	0
Providence	185	74
Richmond	32	2
Scituate	32	6
Smithfield	41	13
South Kingstown	70	0
Tiverton	9	5
Warren	0	0
Warwick	48	8
West Greenwich	16	0

Westerly	13	0
Woonsocket	12	15
Connecticut	4	1
Massachusetts	15	15
Maine	0	1
New Hampshire	2	0
New York	0	2
Regimental Total	954	225

Regimental Statistics

Field and Staff

Died of Disease	0
Died of other causes	0
Mustered Out	14
Died of Disease after mustered out	1
Discharged for Disability	5
Transferred to Veterans Reserve Corps	0
Resigned	2
Dismissed from the service	1
Promoted to other unit	1
Trans. to U.S. Army	0
Trans. to U.S. Navy	0

Combat Casualties

Battle	Killed/Mortally	Wounded	Captured	Total
Fredericksburg	2	3	0	5
Jackson	0	0	1	1
Wilderness	0	0	0	0
Spotsylvania May 12	0	0	0	0
May 13-17	0	0	0	0
May 18	0	0	0	0
North Anna	0	0	0	0
Cold Harbor	0	0	0	0
Bethesda Church	0	0	0	0
Petersburg- initial battles	0	0	0	0
Crater	0	0	0	0
Poplar Spring Church	0	0	0	0
Fort Hell	0	1	0	0
Storming of Petersburg	0	0	0	0
Total	2	4	1	7

Company A

Died of Disease	14
Died of other causes	2
Mustered Out	38
Died of Disease after mustered out	2
Discharged for Disability	21
Transferred to Veterans Reserve Corps	9
Resigned	3
Dismissed from the service	0
Promoted to other unit	0
Trans. to U.S. Army	0
Trans. to U.S. Navy	0
Deserted	7

Combat Casualties

Battle	Killed/Mortally	Wounded	Captured	Total
Fredericksburg	3	21	1	25
Jackson	0	1	0	1
Wilderness	0	0	0	0
Spotsylvania May 12	2	3	0	5
May 13-17	0	0	0	0
May 18	1	1	0	2
North Anna	0	2	0	2
Cold Harbor	0	0	0	0
Bethesda Church	2	3	0	5
Petersburg- initial battles	1	1	0	2
Crater	0	0	0	0
Weldon Railroad	0	0	0	0
Poplar Spring Church	0	0	0	0
Fort Hell	0	0	0	0
Storming of Petersburg	1	1	0	2
Total	10	33	1	44

Company B

Died of Disease	8
Died of other causes	0
Mustered Out	0
Died of Disease after mustered out	3
Discharged for Disability	21
Transferred to Veterans Reserve Corps	7
Resigned	2
Dismissed from the service	0
Promoted to other unit	1
Trans. to U.S. Army	0
Trans. to U.S. Navy	0
Deserted	12

Combat Casualties

Battle	Killed/Mortally	Wounded	Captured	Total
Fredericksburg	5	10	0	15
Jackson	0	2	0	2
Wilderness	0	1	0	1
Spotsylvania May 12	0	1	0	1
May 13-17	2	3	0	5
May 18	0	2	0	2
North Anna	0	0	0	0
Cold Harbor	0	1	0	1
Bethesda Church	0	2	0	2
Petersburg- initial battles	0	2	0	2
Crater	0	1	0	1
Weldon Railroad	0	0	0	0
Poplar Spring Church	1	2	0	3
Fort Hell	0	0	0	0
Storming of Petersburg	0	0	0	0
Total	8	27	0	35

Company C

Died of Disease	9
Died of other causes	0
Mustered Out	59
Died of Disease after mustered out	6
Discharged for Disability	28
Transferred to Veterans Reserve Corps	11
Resigned	1
Dismissed from the service	0
Promoted to other unit	0
Trans. to U.S. Army	0
Trans. to U.S. Navy	0
Deserted	4

Combat Casualties

Battle	Killed/Mortally	Wounded	Captured	Total
Fredericksburg	7	10	0	17
Jackson	0	0	0	0
Wilderness	0	1	0	1
Spotsylvania May 12	0	2	0	2
May 13-17	1	0	0	1
May 18	0	3	0	3
North Anna	0	0	0	0
Cold Harbor	0	1	0	1
Bethesda Church	0	4	0	4
Petersburg- initial battles	0	6	0	6
Crater	0	0	0	0
Weldon Railroad	0	0	0	0
Poplar Spring Church	1	0	0	1
Fort Hell	0	0	0	0
Storming of Petersburg	0	5	0	5
Total	9	32	0	41

Company D

Died of Disease	6
Died of other causes	0
Mustered Out	2
Died of Disease after mustered out	0
Discharged for Disability	16
Transferred to Veterans Reserve Corps	5
Resigned	0
Dismissed from the service	1
Promoted to other unit	0
Trans. to U.S. Army	1
Trans. to U.S. Navy	0
Deserted	7

Combat Casualties

Battle	Killed/Mortally	Wounded	Captured	Total
Fredericksburg	0	11	0	11
Jackson	0	0	1	1
Wilderness	0	0	0	0
Spotsylvania May 12	0	0	0	0
May 13-17	0	0	0	0
May 18	2	0	0	2
North Anna	0	0	0	0
Cold Harbor	0	1	1	2
Bethesda Church	1	3	0	4
Petersburg- initial battles	0	1	0	1
Crater	0	0	0	0
Weldon Railroad	0	0	0	0
Poplar Spring Church	0	1	0	1
Fort Hell	0	0	0	0
Storming of Petersburg	0	0	0	0
Total	3	17	2	22

Company E

Died of Disease	6
Died of other causes	0
Mustered Out	35
Died of Disease after mustered out	2
Discharged for Disability	27
Transferred to Veterans Reserve Corps	5
Resigned	1
Dismissed from the service	0
Promoted to other unit	0
Trans. to U.S. Army	0
Trans. to U.S. Navy	0
Deserted	3

Combat Casualties

Battle	Killed/Mortally	Wounded	Captured	Total
Fredericksburg	5	17	0	22
Jackson	0	1	0	1
Wilderness	0	1	0	1
Spotsylvania May 12	1	3	0	4
May 13-17	0	3	0	3
May 18	0	1	0	1
North Anna	0	1	0	1
Cold Harbor	0	0	2	2
Bethesda Church	3	5	0	8
Petersburg- initial battles	0	1	0	1
Crater	0	0	0	0
Weldon Railroad	0	1	0	1
Poplar Spring Church	0	2	0	2
Fort Hell	0	0	0	0
Storming of Petersburg	0	1	0	1
Total	9	37	2	48

Company F

Died of Disease	10
Died of other causes	0
Mustered Out	32
Died of Disease after mustered out	0
Discharged for Disability	12
Transferred to Veterans Reserve Corps	5
Resigned	2
Dismissed from the service	0
Promoted to other unit	0
Trans. to U.S. Army	0
Trans. to U.S. Navy	0
Deserted	6

Combat Casualties

Battle	Killed/Mortally	Wounded	Captured	Total
Fredericksburg	5	10	0	15
Jackson	0	0	0	0
Wilderness	0	0	0	0
Spotsylvania May 12	0	2	0	2
May 13-17	0	1	0	1
May 18	1	3	0	4
North Anna	1	0	0	1
Cold Harbor	0	3	0	3
Bethesda Church	5	3	0	8
Petersburg- initial battles	0	1	1	2
Crater	0	1	0	1
Weldon Railroad	0	0	0	0
Poplar Spring Church	0	1	0	1
Fort Hell	0	0	0	0
Storming of Petersburg	1	0	0	1
Total	13	25	1	39

Company G

Died of Disease	14
Died of other causes	1
Mustered Out	0
Died of Disease after mustered out	3
Discharged for Disability	17
Transferred to Veterans Reserve Corps	4
Resigned	1
Dismissed from the service	0
Promoted to other unit	0
Trans. to U.S. Army	0
Trans. to U.S. Navy	1
Deserted	0

Combat Casualties

Battle	Killed/Mortally	Wounded	Captured	Total
Fredericksburg	12	14	0	26
Jackson	2	2	0	4
Wilderness	0	0	0	0
Spotsylvania May 12	0	4	0	4
May 13-17	0	2	0	2
May 18	2	2	0	4
North Anna	0	1	0	1
Cold Harbor	1	0	0	1
Bethesda Church	0	0	0	0
Petersburg- initial battles	1	5	0	6
Crater	0	1	0	1
Weldon Railroad	0	0	0	0
Poplar Spring Church	0	0	0	0
Fort Hell	0	0	0	0
Storming of Petersburg	0	0	0	0
Total	18	31	0	49

Company H

Died of Disease	10
Died of other causes	0
Mustered Out	58
Died of Disease after mustered out	0
Discharged for Disability	22
Transferred to Veterans Reserve Corps	11
Resigned	1
Dismissed from the service	0
Promoted to other unit	0
Trans. to U.S. Army	0
Trans. to U.S. Navy	0
Deserted	10

Combat Casualties

Battle	Killed/Mortally	Wounded	Captured	Total
Fredericksburg	5	12	1	18
Jackson	1	2	0	3
Wilderness	0	0	0	0
Spotsylvania May 12	0	4	0	4
May 13-17	2	0	0	2
May 18	3	5	0	8
North Anna	0	0	0	0
Cold Harbor	1	1	0	2
Bethesda Church	0	1	0	1
Petersburg- initial battles	0	6	0	6
Crater	0	0	0	0
Weldon Railroad	0	0	0	0
Poplar Spring Church	0	0	0	0
Fort Hell	0	0	0	0
Storming of Petersburg	0	0	0	0
Total	12	31	1	44

Company I

Died of Disease	7
Died of other causes	0
Mustered Out	66
Died of Disease after mustered out	2
Discharged for Disability	18
Transferred to Veterans Reserve Corps	11
Resigned	4
Dismissed from the service	0
Promoted to other unit	0
Trans. to U.S. Army	0
Trans. to U.S. Navy	0
Deserted	7

Combat Casualties

Battle	Killed/Mortally	Wounded	Captured	Total
Fredericksburg	1	26	1	28
Jackson	0	1	0	1
Wilderness	0	1	0	1
Spotsylvania May 12	0	1	0	1
May 13-17	0	0	0	0
May 18	0	3	0	3
North Anna	1	0	0	1
Cold Harbor	1	0	0	1
Bethesda Church	3	8	0	11
Petersburg- initial battles	2	7	0	9
Crater	0	1	0	1
Weldon Railroad	0	0	0	0
Poplar Spring Church	1	0	0	1
Fort Hell	0	0	0	0
Storming of Petersburg	1	1	0	2
Total	10	49	1	57

Company K

Died of Disease	15
Died of other causes	1
Mustered Out	37
Died of Disease after mustered out	3
Discharged for Disability	21
Transferred to Veterans Reserve Corps	8
Resigned	3
Dismissed from the service	0
Promoted to other unit	3
Trans. to U.S. Army	0
Trans. to U.S. Navy	0
Deserted	2

Combat Casualties

Battle	Killed/Mortally	Wounded	Captured	Total
Fredericksburg	4	11	0	15
Jackson	0	1	0	1
Wilderness	0	0	0	0
Spotsylvania May 12	1	5	0	6
May 13-17	0	1	0	1
May 18	1	1	0	2
North Anna	0	0	0	0
Cold Harbor	1	2	0	3
Bethesda Church	2	0	0	2
Petersburg- initial battles	0	7	0	7
Crater	0	0	0	0
Weldon Railroad	0	0	0	0
Poplar Spring Church	1	2	0	3
Fort Hell	0	0	0	0
Storming of Petersburg	0	0	0	0
Total	10	30	0	40

Company B (New Organization)

Died of Disease	1
Died of other causes	1
Original members mustered out 6/9/65	1
Recruits Mustered out 7/13/65	9
4th RIV Veterans mustered out 6/9/65	14
4th RIV Veterans mustered out 7/13/65	59
Died of Disease after mustered out	1
Discharged for Disability	0
Transferred to Veterans Reserve Corps	0
Resigned	1
Dismissed from the service	0
Promoted to other unit	0
Trans. to U.S. Army	0
Trans. to U.S. Navy	0
Deserted	0

Combat Casualties

Battle	Killed/Mortally	Wounded	Captured	Total
Fort Hell	0	1	0	1
Storming of Petersburg	0	2	0	2
Total	0	3	0	3

Company D (New Organization)

Died of Disease	3
Died of other causes	0
Original members mustered out 6/9/65	0
Recruits mustered out 7/13/65	13
4th RIV Veterans mustered out 6/9/65	12
4th RIV Veterans mustered out 7/13/65	63
Died of Disease after mustered out	2
Discharged for Disability	1
Transferred to Veterans Reserve Corps	1
Resigned	1
Dismissed from the service	0
Promoted to other unit	0
Trans. to U.S. Army	0
Trans. to U.S. Navy	0
Deserted	0

Combat Casualties

Battle	Killed/Mortally	Wounded	Captured	Total
Fort Hell	0	2	0	2
Storming of Petersburg	0	3	0	3
Total	0	5	0	5

Company G (New Organization)

Died of Disease	1
Died of other causes	0
Original members mustered out 6/9/65	0
Recruits mustered out 7/13/65	2
4th RIV Veterans mustered out 6/9/65	17
4th RIV Veterans mustered out 7/13/65	58
Died of Disease after mustered out	2
Discharged for Disability	1
Transferred to Veterans Reserve Corps	0
Resigned	0
Dismissed from the service	0
Promoted to other unit	0
Trans. to U.S. Army	0
Trans. to U.S. Navy	0
Deserted	0

Combat Casualties

Battle	Killed/Mortally	Wounded	Captured	Total
Fort Hell	0	0	0	0
Storming of Petersburg	0	1	0	1
Total	0	1	0	1

REGIMENTAL TOTALS

Cause	Original Members	4th R.I.V. transfers
Died of Disease	99	5
Died of other causes	4	1
Original members mustered out 6/9/65	342	0
Recruits mustered out 7/13/65	24	0
4th RIV Veterans mustered out 6/9/65	0	42
4th RIV Veterans mustered out 7/13/65	0	180
Died of Disease after mustered out	22	5
Discharged for Disability	208	2
Transferred to Veterans Reserve Corps	76	1
Resigned	19	2
Dismissed from the service	2	0
Promoted to other unit	5	0
Trans. to U.S. Army	1	0
Trans. to U.S. Navy	1	0
Deserted	58	0

Combat Casualties

Battle	Killed/Mortally	Wounded	Captured	Total
Fredericksburg	49	145	3	197
Jackson	3	10	2	15
Wilderness	0	4	0	4
Spotsylvania May 12	4	25	0	29
May 13-17	5	10	0	15
May 18	10	21	0	31
North Anna	2	4	0	6
Cold Harbor	4	9	3	16
Bethesda Church	16	29	0	45
Petersburg- initial battles	4	37	1	42
Crater	0	4	0	4

Weldon Railroad	0	1	0	1
Poplar Spring Church	4	8	0	12
Fort Hell	0	1 (3)	0	4
Storming of Petersburg	3	8 (6)	0	17
Total	104	316 (9)	9	438

() indicates members of the Fourth R.I. Volunteers who transferred into the Seventh R.I. Volunteers

FURTHER READING

George H. Allen, *Forty-Six Months with the Fourth R.I. Volunteers in the War of 1861 to 1865: Comprising a History of the Marches, Battles, and Camp Life*. Providence: J.A. & R.A. Reid, 1887.

The diary of a member of Company B of the Fourth Rhode Island who later served in the Seventh Rhode Island. This book was later adopted as the official regimental history of the Fourth Rhode Island and provides excellent detail regarding the unhappy consolidation of the Fourth and Seventh Regiments in the fall of 1864.

John Russell Bartlett, *Memoirs of Rhode Island Officers: Who were Engaged in the Service of their Country During the Great Rebellion of the South*. Providence: Sydney S. Rider, 1867.

A great resource that provides valuable information on the officer corps of the Seventh Rhode Island.

Zenas Randall Bliss, *The Reminiscences of Major General Zenas R. Bliss: 1854-1876*. Edited by Thomas T. Smith, Jerry D. Thompson, Robert Wooster, and Ben E. Pingenot. Austin: Texas State Historical Association, 2007.

The published memoirs of Colonel Bliss of the Seventh, they provide one of the best views of the regiment, particularly their participation at the Battle of Fredericksburg.

John Hutchins Cady,. *Rhode Island Boundaries, 1636-1936*. Providence: Rhode Island Tercentenary Commission, 1936.

Rhode Island's boundaries have shifted tremendously since the Seventh Rhode Island was recruited in 1862. This book is especially helpful in understanding what towns the men enlisted from.

Elisha Dyer, *Annual Report of the Adjutant General of Rhode Island and Providence Plantations, for the Year 1865.* Providence: E.L. Freeman & Son, 1893.

More often cited as the *Revised Register of Rhode Island Volunteers*. Volume One contains the infantry enlistments from Rhode Island and Volume Two contains the men who joined the artillery, cavalry, and Regular Army, as well as the Navy and Marine Corps. This book is a great reference and was the building block to building the roster in this book, however it does contain many errors.

Robert Grandchamp, *Providence to Fort Hell: Letters from Company K, Seventh Rhode Island Volunteers.* Westminster, MD: Heritage Books, 2008.

This book contains the transcribed and edited letters of several members of Company K of the Seventh.

William P. Hopkins, *The Seventh Rhode Island Volunteers in the Civil War:1862-1865.* Providence: Providence Press, 1903.

The official history of the Seventh Rhode Island was written by a drummer in Company D. It is lavishly illustrated with hundreds of images, as well as full of biographical details of members of the regiment. Written in a diary format, it has been cited as among the finest of post-war regimental histories.

Daniel P. Jones, *The Economic and Social transformation of rural Rhode Island, 1780 - 1850.* Boston: Northeastern University Press, 1992.

This book provides an interesting social history of rural western Rhode Island, communities which provided many of the soldiers to the Seventh and is a must read to understand where these men came from.

Proceedings at the Dedication of the Soldiers' and Sailors' Monument, in Providence: To Which Is Appended a List of the Deceased Soldiers and Sailors Whose Names Are Sculptured Upon the Monument. Providence: A.C. Greene, 1871.

The official program for the 1871 dedication of Rhode Island's monument to her Civil War dead in Providence. The book contains a roster of the deceased Rhode Island soldiers and these names were used to cross-reference against the roster in this book.

John E. Sterling, James Lucas Wheaton, and Cherry Fletcher Bamberg. *South Kingstown, Rhode Island Historical Cemeteries.* Greenville: Rhode Island Genealogical Society, 2004.

One of the many fine Rhode Island Civil War books written by John Sterling and others. These books provide fantastic details about where the men of the Seventh Rhode Island are buried and helped to confirm the deaths of several men who died at home after being discharged from the army. As Company G of the Seventh was recruited in South Kingstown, this book was especially helpful.

Edwin W. Stone, *Rhode Island in the Rebellion.* Providence: George H. Whitney, 1865.

Written while the Civil War was still being fought, this book is perhaps the best single volume source for general information about Rhode Island's role in the Civil War.

ACKNOWLEDGEMENTS

At the Rhode Island State Archives, Ken Carlson was instrumental in finding many of the smaller government publications. As always, Kris VanDenBossche pointed me in the path of some of the smaller sources and provided access to his wonderful collection.

In Providence, General Richard Valente provided access to the Benefit Street Arsenal and its vast resources while I was working on the book *Rhody Redlegs*. The staffs at the Rhode Island Historical Society, Providence City Hall Archives, Brown University, and the Providence Public Library were equally helpful.

To the clerks of every city and town hall I visited in Rhode Island, thank you. I am especially indebted to the staff at the halls in Scituate, Glocester, Coventry, Foster, Hopkinton, and Pawtucket.

Captain Phil DiMaria of Battery B has been a mentor, friend, and guide for nearly twenty years as I navigated and researched the role of Rhode Island in the Civil War era. Without Phil's assistance and guidance, none of this work would have been possible.

Nina Wright and the staff at the Westerly Public Library always provided access and many photocopies when I visited that wonderful institution, as did Matt Reardon of the New England Civil War Museum in Rockville, Connecticut.

At the Varnum Continentals, Patrick Donovan provided access to the collections and listened to my many stories.

Midge Frazel helped in ways too important to list.

Cherry Fletcher Bamberg of the Rhode Island Genealogical Society is to be commended to guiding my research and writing over the years as I wrote many articles for *Rhode*

Island Roots. I am also indebted to Rachel Peirce and the other Seventh Rhode Island descendants I have met through the Rhode Island Genealogical Society for providing me information on their ancestors.

Master Sergeant Jim Loffler, the historical section chief of the Rhode Island National Guard was helpful in tracking down the burial locations of some veterans.

I particularly want to thank the staffs at Vicksburg National Military Park, Arlington National Cemetery, Fredericksburg and Spotsylvania National Military Park, Richmond National Battlefield, and Petersburg National Battlefield for providing burial information on Seventh Rhode Island veterans buried there.

Furthermore, I wish to thank the many property owners whose backyards I have visited to locate cemeteries on private property. As well as the Providence Water Supply Board for allowing me to visit the final resting place of my Knight ancestors.

John Sterling and his colleagues who have produced the famed "Rhode Island Cemetery Books," provided an invaluable resource.

Although many years have passed, the interlibrary loan staff and Marlene Lopes at Rhode Island College Special Collections will always be remembered for their assistance in finding long lost books and articles while I was a student there from 2004-2010.

Many of these sources were found in various repositories throughout Rhode Island and although I may not have remembered names, I do wish to thank these institutions that assisted in this work: Langworthy Public Library, East Providence Historical Society, Foster Preservation Society, Scituate Preservation Society, Newport Historical Society, Redwood Library, Burrillville Historical and Preservation Society, Pettaquamscutt Historical Society, Glocester Heritage Society, Bristol Historical Society, North Kingstown Public Library, East

Greenwich Public Library, Westerly Armory Foundation, and the South County Museum.

Lastly, I must thank my dear wife Elizabeth. She has the patience of a saint and gladly lives with the Civil War every day.

ABOUT THE AUTHOR

Robert Grandchamp first became interested in the Seventh Rhode Island Volunteers in 2001, after learning from his grandmother that his third great uncle, Alfred Sheldon Knight had served in the Seventh as a private in Company C and died of pneumonia serving in the Civil War. Robert became very interested in the Civil War in general at this time, and began to read frequently about the Seventh and its campaigns. Trips to battlefields, libraries, and archives fueled his interest and he soon began to collect material for a regimental history that was published in 2008 as *The Seventh Rhode Island Infantry in the Civil War.* Among his other works are *Rhody Redlegs, The Boys of Adams' Battery G, Colonel Edward E. Cross, Rhode Island and the Civil War: Voices from the Ocean State,* and *A Connecticut Yankee at War: The Life and Letters of George Lee Gaskell.* Robert earned his M.A. in American history from Rhode Island College, in addition to his B.A. in anthropology and American history from Rhode Island College as well. He is a former National Park Ranger with service at Shenandoah and Harpers Ferry battlefield. For his efforts to honor the soldiers from Rhode Island, Robert has been awarded the Order of Saint Barbara from the Rhode Island National Guard, the Margaret B. Stillwell Prize from the John Russell Bartlett Society at Brown University, as well as letters of commendation from the governor of Rhode Island and mayor of Providence. Among his professional affiliations, he is a longtime member of several historical organizations, including the Rhode Island Genealogical Society. Robert is an analyst with the Federal government and resides with his wife Elizabeth in Jericho Center, Vermont.

www.ingramcontent.com/pod-product-compliance
Lightning Source LLC
Chambersburg PA
CBHW070730160426
43192CB00009B/1385